Reveal MATH®

Student Practice Book

Mc
Graw
Hill

mheducation.com/prek-12

Send all inquiries to:
McGraw Hill
8787 Orion Place
Columbus, OH 43240

ISBN: 978-0-07-693705-9
MHID: 0-07-693705-4

Printed in the United States of America.

7 8 9 LHS 25 24 23 22

Grade 2
Table of Contents

Unit 2
Place Value to 1,000

Lessons

Unit 3
Patterns within Numbers

Lessons

Unit 4
Meanings of Addition and Subtraction

Lessons

Unit 5

Strategies to Fluently Add within 100

Lessons

Unit 6

Strategies to Fluently Subtract within 100

Lessons

Unit 7
Measure and Compare Lengths

Lessons

Unit 8
Measurement: Money and Time

Lessons

Unit 9
Strategies to Add 3-Digit Numbers

Lessons

Unit 10
Strategies to Subtract 3-Digit Numbers

Lessons

Unit 11
Data Analysis

Lessons

Unit 12
Geometric Shapes and Equal Shares

Lessons

Additional Practice

Name _____

Review

You can group 10 tens to make 1 hundred.

Emma has 10 sheets of stickers. There are 10 stickers on each sheet. How many stickers does Emma have in all?

You can use a tens rod to show each sheet of stickers.

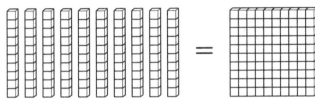

10 tens = **1 hundred**

Emma has 100 stickers in all.

What is the value of the base-ten blocks shown?

1.

_____

2.

_____

3.

_____ tens = _____ hundreds = _____

Use base-ten blocks to show the problem.

4. Jayden read a book for 10 minutes each day for 10 days. How many minutes did Jayden read?

    _____ minutes

5. Sofia uses 10 packs of beads to make 1 bracelet. There are 10 beads in each pack. She made 4 bracelets. How many beads did Sofia use to make the bracelets?

    _____ beads

6. Carlos wants to put 720 of his blocks in his toy box. A small set has 10 blocks and a large set has 100 blocks. How can you write three ways Carlos can put small and large sets of blocks in his toy box?

Math @ Home Activity

Provide opportunities for your child to use groups of ten to make groups of hundreds. For example, have your child place small objects, such as beans, in groups of tens to make 200 objects.

Lesson 2-2

Additional Practice

Name _____

Review

A 3-digit number has hundreds, tens, and ones. Base-ten blocks can be used to represent a 3-digit number. You can use a place-value chart to help you understand the value of the blocks.

What number do these base-ten blocks show?

hundreds	tens	ones
1	3	2

1 flat	3 rods	2 units	The digits show the value of
100	30	2	the base-ten blocks is 132.

What number does each group of base-ten blocks show? Write the number in the place-value chart.

1.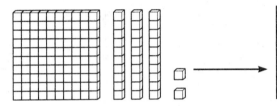

hundreds	tens	ones

2.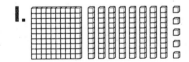

hundreds	tens	ones

What is the value of the 8 in each number?

3. 389: _____

4. 807: _____

What is the value of the digit in the ones place in each number?

5. 431: _____

6. 729: _____

Solve the problem.

7. Ben says the base-ten blocks have a value of 140. Is Ben correct? How do you respond to him?

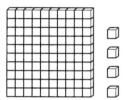

8. Rosa is trying to exercise 175 minutes. She has already walked 110 minutes. She can do jumping jacks for 1 minute at a time. She can jog for 10 minutes at a time. How can Rosa reach her goal? Explain.

Math @ Home Activity

Create a place-value chart that shows hundreds, tens, and ones. Describe a number to your child. Have him or her write numbers on self-sticking notes that he or she will place in the chart to show the hundreds, tens, and ones in your number. Then switch roles and repeat the activity.

Additional Practice

Name _____

Review

You can read and write 3-digit numbers using place value, words, and numerals.

Expanded Form

The base-ten blocks show 214.

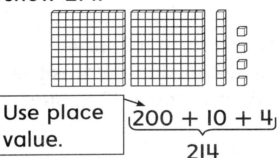

| Use place value. | → 200 + 10 + 4 |

214

Word Form

200 + 10 + 4

two hundred fourteen

← | Use words. |

Standard Form

214 ← | Use numerals. |

How can you write the number in standard form?

I. one hundred one _____

2. three hundred twenty-five _____

3. five hundred sixty-two _____

How can you write the number in expanded form?

4. 236   _____ + _____ + _____

5. 466   _____ + _____ + _____

6. 784   _____ + _____ + _____

7. Write the standard form, expanded form, and word form for the value of the base-ten blocks.

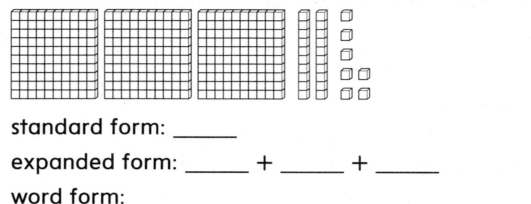

standard form: _____

expanded form: _____ + _____ + _____

word form: _____

8. Antonio has one hundred twelve rocks. How can he show the number of rocks in standard form?

_____ rocks

9. Landon writes the standard form for three hundred twenty-six as 300206. Jane writes the standard form as 326. Who is correct? Explain.

10. Kate wrote the word form and expanded form of 725. How do you respond to her?

word form: seven hundred twenty-five
expanded form: 7 + 2 + 5

Copyright © McGraw-Hill Education

Lesson 2-4

Additional Practice

Name _____

Review

You can decompose a 3-digit number by grouping the hundreds, tens, and ones in different ways by place value.

One Way

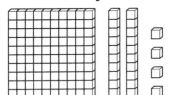

I hundred, 2 tens, 4 ones

Another Way

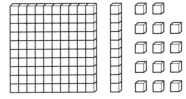

I hundred I ten 14 ones

---

1. Show 352 decomposed in two different ways.

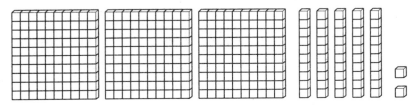

_____ hundreds, _____ tens, _____ ones

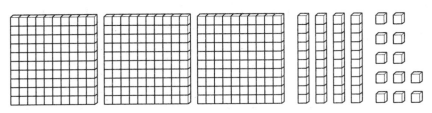

_____ hundreds, _____ tens, _____ ones

How can you decompose the number? Choose all the correct answers.

2. 567

    A. $5 + 6 + 7$

    B. $500 + 60 + 7$

    C. $500 + 50 + 17$

    D. $567 + 67 + 7$

3. 839

    A. $800 + 3 + 9$

    B. $800 + 30 + 9$

    C. $800 + 30 + 19$

    D. $800 + 20 + 19$

Decompose the number in two ways.

4. _____ + _____ + _____ = 758

    _____ + _____ + _____ = 758

5. _____ + _____ + _____ = 924

    _____ + _____ + _____ = 924

---

6. How can you decompose 132 into tens and ones? Explain.

7. James says he can decompose 416 into 3 hundreds, 11 tens, and 6 ones. How do you respond to him?

Math @ Home Activity

Write a 3-digit number on a piece of paper. Have your child draw and cut out base-ten blocks to show how to decompose the number in two different ways. For example, your child could show how to decompose 235 in two different ways by drawing and cutting out 2 hundreds flats, 3 tens rods, and 5 ones units and 2 hundreds flats, 2 tens rods, and 15 ones units. Repeat the activity with a different 3-digit number.

Additional Practice

Name _____

Review

You can use place value to compare 3-digit numbers.

Compare the values of the hundreds first.	If the hundreds have the same value, compare the values of the tens.

hundreds	tens	ones
6	4	1
5	4	7

hundreds	tens	ones
5	4	7
5	8	9

600 is **greater than** 500

So, 641 > 547

40 is **less than** 80.

So, 547 < 589

How can you compare the numbers? Use >, <, or =.

1.
hundreds	tens	ones
8	8	0
8	0	8

880 ◯ 808

2.
hundreds	tens	ones
4	4	7
4	7	4

447 ◯ 474

How can you compare the numbers? Use >, <, or =.

3. 155 ◯ 317

4. 690 ◯ 609

5. 298 ◯ 297

6. 788 ◯ 788

7. 521 ◯ 525

8. 801 ◯ 811

Circle the number *greater than* the number in the box.

9. | 613 | 612   614

10. | 941 | 944   914

---

11. A number is greater than 3 hundreds, 8 tens, and 6 ones. The number is less than 3 hundreds, 8 tens, and 8 ones. What is the number?

_____

12. There are 112 second graders and 120 third graders at a school. Which grade has a greater number of students? Explain how you know.

Math @ Home Activity

Create a flash card for each of the following symbols: >, <, and =. Write two 3-digit numbers on separate pieces of paper, hold them up, and have your child hold up a symbol card to correctly compare the numbers. Repeat the activity several times.

Additional Practice

Name

Review

You can look for place-value patterns to help you count.

101	102	103	104	105	106	107	108	109	110
111	112	113	114	115	116	117	118	119	120
121	122	123	124	125	126	127	128	129	130
131	132	133	134	135	136	137	138	139	140
141	142	143	144	145	146	147	148	149	150
151	152	153	154	155	156	157	158	159	160
161	162	163	164	165	166	167	168	169	170
171	172	173	174	175	176	177	178	179	180
181	182	183	184	185	186	187	188	189	190
191	192	193	194	195	196	197	198	199	200

The ones digits go up by 1 from left to right in each row.

The ones digit changes to 0 and the tens digit goes up by 1.

The ones digit and tens digit change to 0 and the hundreds digit goes up by 1.

The tens digits go up by 1 from top to bottom in each column.

---

1. What numbers are missing? Fill in the blanks.

501	502	503	504	505		507	508		510
	512	513	514		516	517		519	
521			524	525			528		530

What number is missing? Fill in the blank.

2. 345, 346, 347, _____   3. 719, 720, 721, _____

4. 218, 219, _____, 221   5. 577, 578, _____, 580

What numbers are missing? Fill in the blanks.

6. _____, 209, 210, _____

7. 464, _____, 466, _____

8. _____, _____, 650, 651

9. 981, 982, _____, _____

10. 858, _____, 860, _____

---

11. Edmund is writing numbers from 1 to 1,000. He writes 989, 990, 991, and 992. What are the next 5 numbers he will write? Explain your thinking.

Math @ Home Activity

Work with your child to develop counting by ones. Beginning with a number from 0 to 990, count aloud to your child four sequential numbers. Then have your child say the next 5 numbers. For example, if you say 150, 151, 152, 153, your child says 154, 155, 156, 157, 158. Repeat with additional numbers.

Additional Practice

Name _____

Review

You can use a number chart or a number line to skip count by 5s.

You can skip count by 5s from a number ending in 0 or 5.

1	2	3	4	**5**	6	7	8	9	**10**
11	12	13	14	**15**	16	17	18	19	**20**
21	22	23	24	**25**	26	27	28	29	**30**
31	32	33	34	**35**	36	37	38	39	**40**
41	42	43	44	**45**	46	47	48	49	**50**
51	52	53	54	**55**	56	57	58	59	**60**
61	62	63	64	**65**	66	67	68	69	**70**
71	72	73	74	**75**	76	77	78	79	**80**
81	82	83	84	**85**	86	87	88	89	**90**
91	92	93	94	**95**	96	97	98	99	**100**

The ones digits are 5 or 0.

You can also skip count by 5s from a number *not* ending in 0 or 5.

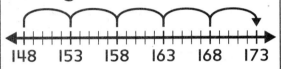

148   153   158   163   168   173

The ones digits alternate between two numbers that are 5 apart.

---

1. Use the number chart to skip count.

Start at 4. Count by 5s. Color the numbers.

What do you notice?

1	2	3	4	5	6	7	8	9	10
11	12	13	14	15	16	17	18	19	20
21	22	23	24	25	26	27	28	29	30
31	32	33	34	35	36	37	38	39	40
41	42	43	44	45	46	47	48	49	50
51	52	53	54	55	56	57	58	59	60
61	62	63	64	65	66	67	68	69	70
71	72	73	74	75	76	77	78	79	80
81	82	83	84	85	86	87	88	89	90
91	92	93	94	95	96	97	98	99	100

How can you skip count by 5s? Fill in the number.

2. 200, 205, 210, _____   3. 545, 550, 555, _____

4. 626, 631, 636, _____   5. 977, 982, 987, _____

How can you count by 5s? Fill in the numbers.

6.
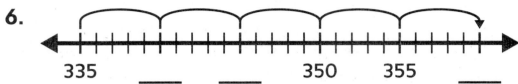

335 _____ _____ 350 355 _____

7.

_____ 729 _____ _____ _____ 749

---

8. Ethan has placed 162 photos in an album. If he places 5 more groups of 5 photos in the album, how many photos will he have placed in the album? Explain your thinking.

Math @ Home Activity

Create word problems about everyday situations requiring your child to skip count by 5s. For example, identify the number of school days your child has attended and count by 5s to find how many days they will have attended by the end of the next 4 weeks. Find or create a number chart for your child to use to explain his or her answers.

Additional Practice

Name _____

Review

You can notice patterns when you skip count by 10s and 100s.

10	20	30	40	50	60	70	80	90	**100**
110	120	130	140	150	160	170	180	190	**200**
210	**220**	**230**	**240**	250	260	270	280	290	**300**

When skip counting by 10s, the tens digit goes up by 1.

When skip counting by 100s, the hundreds digit goes up by 1.

328   338   348   358   619   719   819   919

What value is shown by each jump? Fill in the blanks.

1.

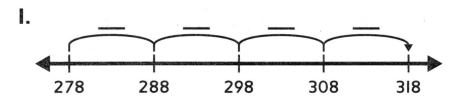

278   288   298   308   318

2.

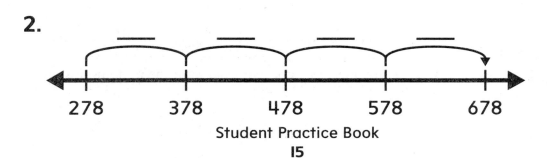

278   378   478   578   678

3. How can you skip count to find the missing numbers? Fill in the blanks.

510		530	540	550	560	570		590	
	620	630		650	660		680	690	700
710	720		740			770	780	790	

How can you skip count on a number line? Fill in the numbers.

4. Skip count by 10s.

___   412   ___   ___   ___

5. Skip count by 100s.

___   ___   ___   ___   956

---

6. What is the pattern when you skip count by 10s? What is the skip counting by 100s pattern?

Math @ Home Activity

Draw a 10-by-10 grid. Have your child create a number chart by writing 0 in the top left square and then skip count by 10s to fill in the rest of the squares. Then ask your child what the patterns are when skip counting by 10s and by 100s.

Additional Practice

Name _____

Review

You can pair objects from a group or skip count by 2s to determine even and odd numbers.

Feng has 12 books. Does he have an even or odd number of books?

An even number of books can be paired with no books left.

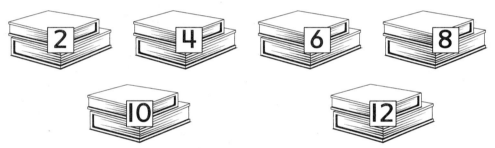

The number 12 is even. Feng has an even number of books.

Is the number Even or Odd? Draw to show your thinking. Circle the answer.

1. ⬚ 5   Even   Odd

2. ⬚ 8   Even   Odd

3. ⬚ 10   Even   Odd

4. ⬚ 17   Even   Odd

5. Circle all the odd numbers on the number chart. Then cross out all the even numbers on the number chart.

1	2	3	4	5	6	7	8	9	10
11	12	13	14	15	16	17	18	19	20

6. Jalen buys 16 markers. Does he buy an odd or even number of markers? Draw to show your thinking.

7. There is an even number of toys in a toy box. The number of toys is between 2 and 5. How many toys are in the toy box? Explain how you know.

8. Carmen has an odd number of stamps. The number of stamps is between 15 and 18. How many stamps does Carmen have? Explain how you know.

Math @ Home Activity

Write the numbers 1 through 20 on separate index cards. Shuffle the cards and place the deck facedown. Have your child turn one card over, draw objects to represent the number, and state whether the number is even or odd. Encourage your child to repeat the activity with a different card until all cards have been used.

Additional Practice

Name _____

Review

You can write an even number with a doubles fact.

Even numbers can be separated into two equal groups.

$$4 + 4 = 8$$

● ● ○ ○
● ● ○ ○

8 is an even number.

Odd numbers *cannot* be separated into two equal groups.

$$4 + 5 = 9$$

● ● ○ ○ ○
● ● ○ ○

9 is an odd number.

Show the even number as the sum of a doubles fact.

1. 2 = ____ + ____

2. 12 = ____ + ____

3. 10 = ____ + ____

4. 6 = ____ + ____

5. Write two equations with even sums.

____ + ____ = ____

____ + ____ = ____

6. Write two equations with odd sums.

____ + ____ = ____

____ + ____ = ____

---

7. Gemma is sharing her crayons with her sister. Can the crayons be separated equally between Gemma and her sister? How do you know?

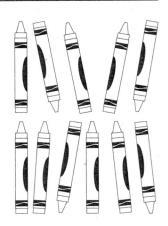

8. Jake has 6 red marbles and 8 orange marbles. Is his total number of marbles an even or odd number? Explain how you know.

Math @ Home Activity

Find 20 small objects, such as pennies or buttons. Choose a number from 2 to 20. Have your child use the objects to show the number. Then ask your child to write an equation to show the number. Repeat this activity several times with different numbers of small objects.

Additional Practice

Name _____

Review

You can skip count the amount in each row to find the total number of objects in an array.

How can Juan find how many toy cars he has?

Arrange the toy cars in 3 rows of 4 toy cars.

Skip count each row by 4s.

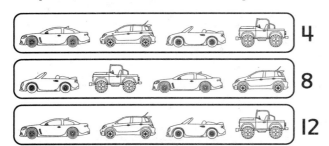

4

8

12

There are 12 toy cars in the array.

Skip count to find the number of objects in the array.

1.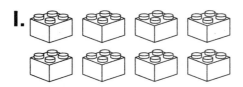

_____ building blocks

2.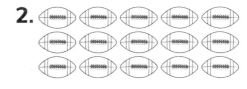

_____ footballs

3. Fill in the numbers to skip count the number of dolls in each column.

Sara displays her dolls on shelves. How many dolls does Sara have?

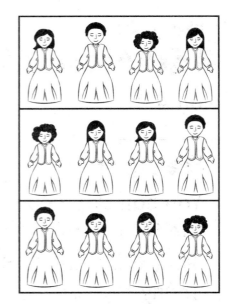

3, _____, _____, _____

_____ dolls

---

4. How can you skip count to find the number of counters in the array? Choose the correct answer.

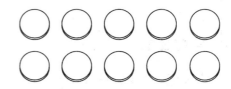

 A. 2, 4   **B.** 2, 4, 6

 C. 2, 4, 6, 8   **D.** 2, 4, 6, 8, 10

---

5. Carl has 7 trophies. Can Carl arrange his trophies in an array? Explain your thinking.

Math @ Home Activity

Have your child use small everyday objects, such as coins or beans, to create an array with up to 5 rows and 5 columns. Then have your child skip count to determine the total number of objects in the array. Repeat several times.

Additional Practice

Name _____

Review

You can make an array to represent a problem and use repeated addition to solve it.

There are 2 rows of 3 carrot plants in a garden. How many carrot plants are there?

○ ○ ○ 3
○ ○ ○ 3

○ ○ ○
○ ○ ○
2 2 2

Add the number in each row: $3 + 3 = 6$

Add the number in each column: $2 + 2 + 2 = 6$

There are 6 carrot plants.

Write two equations to show the array.

1.

_____ + _____ + _____ + _____ = _____

_____ + _____ + _____ = _____

Shade the grid to show the array. Write an equation to describe the array.

2. Show 5 rows and 3 columns.

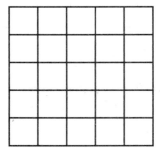

_____

3. Show 5 rows and 5 columns.

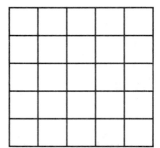

_____

4. Roy places cards in 4 rows and 5 columns. Draw an array to show how many cards Roy has in all. Write an equation to describe the array.

Math @ Home Activity

Make a 5-by-5 array, or outline one on graph paper. Look for an array of objects around your home with 5 to 20 items, such as an egg carton with 2 rows of 4 eggs. Have your child shade the array you made to represent the array of objects. Then have your child write a matching equation. Repeat with different arrays of household objects.

Additional Practice

Name _____

Review

You can use addition to represent a problem in which a number is added to another number.

There are some sheets of paper on a desk. Peter puts 5 more sheets on the desk. Now there is a total of 23 sheets of paper on the desk. How many sheets of paper were on the desk before?

Represent the problem using a part-part-whole mat.

Write an equation to help you find the **unknown addend.**

Part	Part
?	**5**
Whole	
23	

$$? + 5 = 23$$

$$18 + 5 = 23$$

There were **18** sheets of paper on the desk before.

I. Which equation represents the word problem? Choose the correct answer.

Some students are in the computer lab. Then 8 more students go into the computer lab. There are now 25 students in the computer lab. How many students were in the computer lab before?

A. $33 - 8 = ?$ B. $? + 8 = 25$

C. $33 - ? = 25$ D. $8 + 25 = ?$

2. What equation can represent the problem? Solve and explain how your equation relates to the problem.

   A basketball team scores some points. Then they score 6 more points. Now they have 68 points. How many points did they score before?

3. Write a word problem that could be represented by this part-part-whole mat. Write an equation to represent your problem. Use the equation to solve your problem.

Part	Part
?	7
Whole	
34	

Math @ Home Activity

Look for a situation around your home where the first addend in an addition problem is unknown. Ask your child to find the unknown to solve the problem. Have him or her explain the solution by drawing a part-part-whole diagram. Repeat with another situation.

Additional Practice

Name _____

Review

You can use subtraction to represent a problem in which a number is taken from another number.

May opens a box of raisins. She gives her brother 9 raisins. There are 13 raisins left. How many raisins were in the box?

Represent the problem using a bar diagram.

Write an equation to help you find the **unknown**.

?	
9	13

$$? - 9 = 13$$
$$22 - 9 = 13$$

There were **22** raisins in the box.

I. Which equation represents the word problem? Choose the correct answer.

There are some leaves on a tree. 14 leaves fall off. Now there are 21 leaves on the tree. How many leaves were on the tree before?

A. $? + 14 = 21$ **B.** $21 - 14 = ?$

C. $21 - ? = 14$ **D.** $? - 14 = 21$

2. What equation can represent the problem? Solve and explain how your equation relates to the problem.

There are 65 coins in a jar. Ivan takes out 8 coins. How many coins are in the jar now?

3. a. Write a subtraction word problem with an unknown change number.

b. Write an equation that represents your problem. Use the equation to solve your problem.

Math @ Home Activity

Create riddles for your child to solve. The riddles should require your child to find an unknown number that is being subtracted from another number. For example, "There are 22 ants on a log. Some ants crawl away. Now there are 12 ants on the log. How many ants crawled away?"

Additional Practice

Name _____

Review

You can solve problems with more than one step that include addition, subtraction, or both.

Lynn has 12 marbles. She loses 6 marbles. Her sister gives her 3 marbles. Jay has 15 marbles. He buys 8 marbles. He gives 7 marbles to a friend. How many marbles do they each have now?

Draw to represent the problem.

Lynn ((12) − (6) + (3))   Jay ((15) + (8) − (7))

Write equations to help you find the **unknowns**.

$$12 - 6 + 3 = ?$$
$$12 - 6 + 3 = 9$$
Lynn has **9** marbles now.

$$15 + 8 - 7 = ?$$
$$15 + 8 - 7 = 16$$
Jay has **16** marbles now.

---

1. Which steps represent the word problem?

   Aliya has 14 daisies. She gives 6 daisies to her mom. Then she picks 8 daisies. How many daisies does Aliya have now?

   A. Add 14 and 6. Then add 8 to the sum.

   B. Add 14 and 8. Then add 6 to the sum.

   C. Subtract 8 from 14. Then add 6 to the difference.

   D. Subtract 6 from 14. Then add 8 to the difference.

2. What equation can represent the problem? Solve and explain how your equation relates to the problem.

   Darla makes 18 necklaces. She sells 9 of the necklaces. Then she makes 4 necklaces. How many necklaces does Darla have now?

3. Luca has 15 trading cards. He gives his friend 3 trading cards and then gets 5 from his friend. Luca writes the equation $15 + 3 + 5 = ?$ when he tries to find how many trading cards he has now. How do you respond to Luca?

Math @ Home Activity

Create a two-step word problem that involves addition and subtraction for your child to solve. Have him or her write an equation to represent the problem. Then ask your child to explain how the equation relates to the problem and then solve it.

Additional Practice

Name _____

Review

You can use addition or subtraction to represent a problem in which two numbers are put together.

There 16 flowers in a vase. 9 of the flowers are red. The rest of the flowers are white. How many white flowers are in the vase?

Represent the problem using a part-part-whole mat.

Write an equation to help you find the **unknown**.

Part	Part
9	?
Whole	
16	

$$9 + ? = 16 \qquad 16 - 9 = ?$$
$$9 + 7 = 16 \qquad 16 - 9 = 7$$

There are **7** white flowers in the vase.

---

1. Which equations can represent the word problem? Choose all the correct answers.

   Eduardo finds 17 rocks. 6 are small and the rest are large. How many large rocks does Eduardo find?

   A. $17 + 6 = ?$ **B.** $17 - 6 = ?$

   C. $? + 6 = 17$ **D.** $6 + ? = 17$

2. What equation can represent the problem? Solve and explain how your equation relates to the problem.

There are 50 fish in a fish tank. 30 fish are orange. The rest of the fish are blue. How many blue fish are in the tank?

3. Write a word problem that could be represented by the part-part-whole mat. Write an equation that represents your problem. Use the equation to solve your problem.

Part	Part
10	?
Whole	
25	

Math @ Home Activity

Draw a part-part-whole diagram with your child. In the diagram, have him or her write numbers and a question mark to represent the problem, "There are 22 students in Mr. Parker's music class. 10 students play the flute and the rest play a different instrument. How many students don't play the flute?" Next, tell your child to write two different equations that can represent this problem. Finally, have your child solve the problem.

Additional Practice

Name _____

Review

You can use addition or subtraction to represent a problem in which a total is broken into two groups.

Irma has 16 photos. There are 9 photos in a round album and the rest are in a square album. How many photos are in the square album?

Draw to represent the problem.

Write an equation to find the unknown.

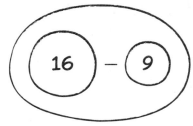

$$16 - 9 = ? \qquad 9 + ? = 16$$
$$16 - 9 = 7 \qquad 9 + 7 = 16$$

There are **7** photos in the square album.

I. Which equations can represent the word problem? Choose all the correct answers.

Wade has 13 toy trucks. 7 trucks are red and the rest are blue. How many trucks are blue?

A. $7 + ? = 13$ **B.** $13 - 7 = ?$

C. $? - 7 = 13$ **D.** $13 + 7 = ?$

2. What equation can represent the problem? Solve and explain how your equation relates to the problem.

A farmer has 18 animals. 9 of the animals are chickens and the rest are pigs. How many animals are pigs?

3. **a.** Write a word problem that has an unknown addend.

b. Use an equation to solve your word problem.

Additional Practice

Name _____

Review

You can solve problems with more than one step that include addition, subtraction, or both.

Carlos sees 10 animals at a park. He sees 5 birds and 3 squirrels. The rest are dogs. How many animals are dogs?

Draw to represent the problem.

Write equations to help you find the **unknown.**

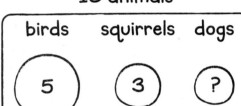

10 animals

$$5 + 3 + ? = 10$$

$$5 + 3 = 8 \text{ and } 8 + 2 = 10$$

2 animals are dogs.

---

1. Write an equation to represent the problem using ? for the unknown. Then solve.

   Everett made 6 towers for his sandcastle. A wave washes 3 towers away. He makes more towers. Now he has 8 towers. How many towers did Everett make?

   a. Equation: _____

   b. Solve: _____ _____

2. What equation can represent the problem? Solve and explain how your equation relates to the problem.

Amy uses 7 beads to decorate a picture frame. She uses 1 blue bead and 2 yellow beads. The rest are green. How many are green?

3. Ben has 9 action figures. He loses 2 of them. He gets some action figures for his birthday. Now he has 10 action figures. Can Ben use the equation $9 - 2 + 10 = ?$ to find how many action figures he has now? Explain.

Math @ Home Activity

Have your child write a two-step word problem that involves addition and subtraction. The problem should mimic the problems in this lesson (a total is given, and a missing addend needs to be found). Help your child as needed. Ask him or her to write an equation to represent the problem and solve it.

Additional Practice

Name _____

Review

You can use addition to represent a compare problem.

Lewis has 4 fewer crayons than Tamara. Lewis has 9 crayons. How many crayons does Tamara have?

Represent the problem using a bar diagram.

Write an equation to help you find the **unknown**.

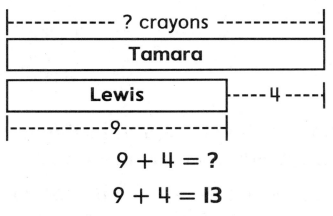

$$9 + 4 = ?$$

$$9 + 4 = 13$$

Tamara has **13** crayons.

---

1. Which equation represents the word problem? Choose the correct answer.

   Will checks out 3 fewer library books than Franco. Will checks out 6 library books. How many library books does Franco check out?

   A. $6 - 3 = ?$ **B.** $6 - ? = 3$

   C. $6 + 3 = ?$ **D.** $? + 3 = 6$

2. Mae colors 11 pages of a coloring book. Avi colors 7 pages of a coloring book. How many more pages does Mae color?

_____ _____

3. Jordan has 15 seashells. Jordan has 2 fewer seashells than Jack. How many seashells does Jack have?

_____ _____

4. Write an equation to represent the problem using ? for the unknown. Then explain how to solve.

Jamie sees 14 butterflies in a flower garden. Jamie sees 5 fewer butterflies than Maya. How many butterflies does Maya see in the flower garden?

Math @ Home Activity

Help your child practice solving compare problems with the greater quantity unknown. Give him or her a group of objects, such as coins, paper clips, or marbles. Then tell your child you have more of the objects than he or she does. Tell your child how many more objects you have. Finally, have him or her determine how many objects you have.

Additional Practice

Name _____

Review

You can use addition or subtraction to represent a compare problem.

Bernice swims 6 more laps than Susie. Bernice swims 17 laps. How many laps does Susie swim?

Represent the problem using a bar diagram.

Use mental math or an equation to find the **unknown**.

$$? + 6 = 17 \qquad 17 - 6 = ?$$
$$11 + 6 = 17 \qquad 17 - 6 = 11$$

Susie swims **11** laps.

---

1. Which equations can represent the word problem? Choose all the correct answers.

   Jamal eats 7 more green beans than Nolan. Jamal eats 14 green beans. How many green beans does Nolan eat?

   A. $14 + 7 = ?$ **B.** $14 - ? = 7$

   C. $7 + ? = 14$ **D.** $? - 7 = 14$

2. What equation can represent the problem? Solve and explain how your equation relates to the problem.

Harmen washes 5 fewer dishes than Lori. Lori washes 13 dishes. How many dishes does Harmen wash?

3. Write a word problem that could be represented by the bar diagram. Write an equation that represents your problem. Then use your equation to solve the problem.

```
|-----------18 flowers-----------|
|                                |
|            Iris                |
|                                |
|        Nelson        |--- 4 ---|
|                      |
|-----------?-----------|
```

Copyright © McGraw-Hill Education

Additional Practice

Name _____

Review

You can solve problems with more than one step that include a comparison.

There are 3 children on slides. There are 5 more children on swings than on slides. How many children are on slides and swings?

Draw to represent the problem.

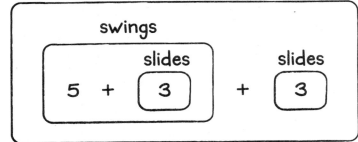

? children on slides and swings

Write equations to help you find the **unknown**.

$5 + 3 + 3 = ?$

$5 + 3 = 8$ and $8 + 3 = 11$

11 children are on slides and swings.

I. Write an equation to represent the steps of the problem. Then solve the problem.

Vance reads 8 more books over the summer than his brother. Vance's brother reads 6 books. How many books did they both read?

Step I equation: _____

Step 2 equation: _____

_____ _____

2. What equation can represent the problem? Solve and explain how your equation relates to the problem.

Thomas buys 2 more apples than bananas. He buys 4 bananas. How many pieces of fruit does Thomas buy?

3. a. Write a word problem with more than one step that includes a comparison.

b. Use equations to solve your word problem.

Look for opportunities around your home to encourage your child to practice solving two-step addition comparison problems. For example, if your child has 3 more red marbles than blue marbles, and he or she has 4 blue marbles, have your child find the total number of red and blue marbles.

Additional Practice

Name _____

Review

You can draw a picture to represent problems with more than one step that include addition, subtraction, or both. Then an equation can be used to solve these problems.

Yi makes 9 sandwiches. His family eats 5 sandwiches. He makes 3 more sandwiches. How many sandwiches are there now?

Draw to represent the problem.

makes   eats   more

9 — 5 + 3

Use mental math or an equation to find the unknown.

$$9 - 5 + 3 = ?$$

$9 - 5 = ?$ $4 + 3 = ?$

$9 - 5 = 4$ $4 + 3 = 7$

There are **7** sandwiches now.

---

1. Which equation represents the word problem? Choose the correct answer.

   Darron earns $8 on Monday and $4 on Tuesday. He buys a toy for $9 on Wednesday. How much money does Darron have left?

   A. $8 + 4 + 9 = ?$ **B.** $8 - 4 + 9 = ?$

   C. $8 - 4 + ? = 9$ **D.** $8 + 4 - 9 = ?$

2. How can you solve the problem? Write the steps. Then solve.

An art teacher places 10 paintbrushes on a table. Some students take 7 paintbrushes. The art teacher puts 5 more paintbrushes on the table. How many paintbrushes are on the table now?

_____ _____

3. There are 12 boxes of yogurt tubes on a grocery store shelf. Customers buy 11 of the boxes. A worker puts 6 more boxes on the shelf. Can the equation $12 + 11 + 6 = ?$ be used to find how many boxes of yogurt tubes are on the shelf now? Explain.

Math @ Home Activity

Look for situations around your home where your child can solve two-step word problems with addition and subtraction. Have your child write and solve equations to reflect the situations. For example, suppose your child is given 8 pretzels. Then he or she eats 5 pretzels. After that, your child is given 4 more pretzels. Ask your child how many pretzels he or she has now.

Additional Practice

Name _____

Review

You can use strategies to fluently add within 20.

$7 + 6 = ?$

One strategy to find the sum is to count on.

0 1 2 3 4 5 6 7 8 9 10 11 12 13 14 15 16 17 18 19 20

Another strategy to find the sum is to decompose one addend to make a 10.

$7 + 6 = ?$

$3 + 3$

$7 + 3 = 10$ $10 + 3 = 13$

How can you decompose the second addend to make a 10? Circle the correct answer.

1. $6 + 5 = ?$

$3 + 2$ $4 + 1$

2. $7 + 4 = ?$

$3 + 1$ $2 + 2$

3. $8 + 6 = ?$

$2 + 4$ $3 + 3$

4. $9 + 7 = ?$

$2 + 5$ $1 + 6$

What is the sum? Use the number line to solve.

5. 5 + 7 = _____

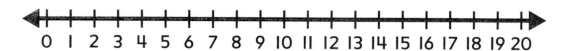

6. 8 + 9 = _____

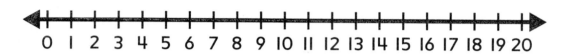

7. 9 + 4 = _____

---

8. Rachel has 9 grapes. Her mother gives her 7 more grapes. What is the total number of grapes Rachel has?

    _____

9. Juan is trying to solve 8 + 5 = ?. How can he decompose 5 to make 10 in the equation? Explain.

Math @ Home Activity

On a sheet of paper, write an addition expression with a sum less than 20, for example, 7 + 5 = ?. Have your child use small household objects, such as paper clips or beans, to represent making a 10 to find the sum. Repeat this activity several times.

Additional Practice

Name _____

Review

You can use doubles facts to help find the sums of near doubles facts to fluently add within 20.

$7 + 8 = ?$ | $7 + 9 = ?$

$7 + 8$ is a doubles + 1 fact. | $7 + 9$ is a doubles + 2 fact.

$7 + 8 = 15$ | $7 + 9 = 16$

$7 + 1$ | $7 + 2$

$14 + 1 = 15$ | $14 + 2 = 16$

So, $7 + 8 = 15$. | So, $7 + 9 = 16$.

How can you decompose the second addend to make a double with the first addend? Circle the correct answer.

1. $4 + 5 = ?$

   $4 + 1$ $3 + 2$

2. $6 + 7 = ?$

   $4 + 3$ $6 + 1$

3. $3 + 5 = ?$

   $1 + 4$ $3 + 2$

4. $6 + 8 = ?$

   $6 + 2$ $5 + 3$

How can you use doubles to solve? Complete the equation.

5.

_____ + _____ = _____

6.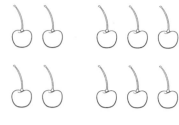

_____ + _____ = _____

7.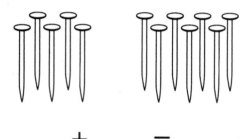

_____ + _____ = _____

8.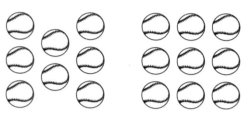

_____ + _____ = _____

---

9. There are 6 bananas in one bunch and 8 bananas in another bunch. How can you decompose 8 to use a doubles fact to find the total number of bananas? Explain.

Math @ Home Activity

Use small objects, such as macaroni, to represent a doubles fact. Then have your child write the related doubles + 1 and doubles + 2 facts. If necessary, have your child also use the small objects to determine the related facts.

Additional Practice

Name _____

Review

You can use base-ten blocks to help you add 2-digit numbers.

Show each number using base-ten blocks.
Add the ones. Then add the tens.
Regroup 10 ones as 1 ten, if needed.

$35 + 24 = ?$

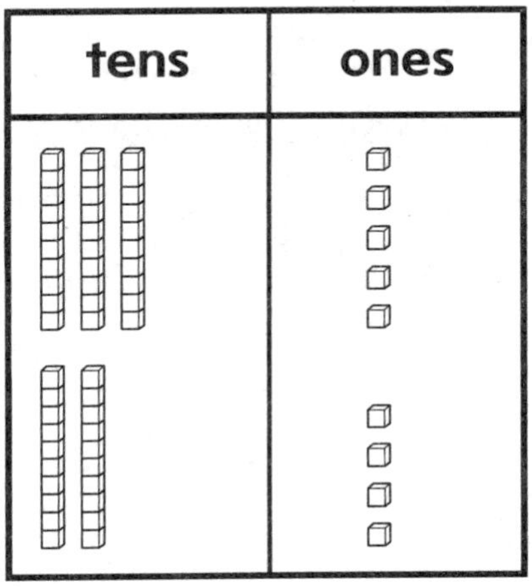

tens	ones

$35 + 24 = 59$

$56 + 37 = ?$

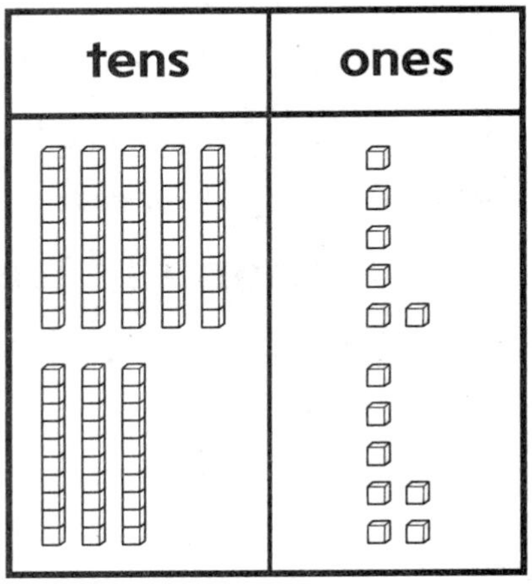

tens	ones

$56 + 37 = 93$

In which equation do 10 ones need to be regrouped? Circle your answer.

1. $43 + 27 = ?$

   $58 + 31 = ?$

2. $52 + 23 = ?$

   $36 + 45 = ?$

3. $38 + 43 = ?$

   $62 + 36 = ?$

4. $25 + 45 = ?$

   $28 + 71 = ?$

What is the sum? Use base-ten shorthand or base-ten blocks to show your thinking.

5. 41 + 26 = _____

tens	ones

6. 53 + 39 = _____

tens	ones

---

7. Marnie has 54 sheets of construction paper. Her mother gives her 28 more sheets. Marnie says she has 82 sheets of construction paper. How do you respond to Marnie? Use base-ten shorthand to explain your thinking.

Math @ Home Activity

Ask your child to add 2-digit numbers to find sums less than 100 during everyday situations at home. For example, as you watch a basketball game, ask your child to determine how many points two players scoring more than 10 points each scored together. Encourage him or her to use base-ten shorthand to find the sum.

Additional Practice

Name _____

Review

You can add addends in any order.

$23 + 35 = ?$

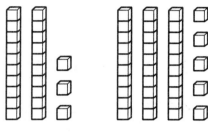

$23 + 35 = 58$

$35 + 23 = ?$

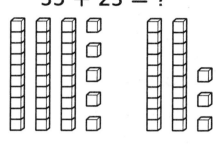

$35 + 23 = 58$

Complete the equations shown by the base-ten blocks.

1.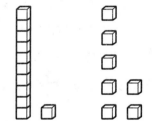

   $11 + 7 =$ _____

   $7 +$ _____ $=$ _____

2.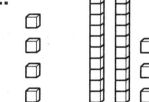

   _____ $+ 23 = 27$

   _____ $+ 4 =$ _____

3.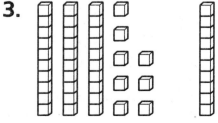

   $38 +$ _____ $= 48$

   $10 +$ _____ $=$ _____

4.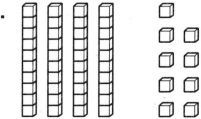

   _____ $+ 9 = 49$

   _____ $+ 40 =$ _____

What is the sum?

5. 44 + 35 = _____

    35 + 44 = _____

6. 27 + 14 = _____

    14 + 27 = _____

7. 42 + 18 = _____

    18 + 42 = _____

8. 36 + 39 = _____

    39 + 36 = _____

9. 15 + 17 = _____

    17 + 15 = _____

10. 74 + 21 = _____

    21 + 74 = _____

---

11. Riley collects 18 leaves. Then Riley goes out and collects 22 more leaves. Riley says the only way she can find her total number of leaves is to add 18 + 22, because she had 18 leaves first. How do you respond to Riley?

Math @ Home Activity

Place 22 objects on the left side of a tray and 13 objects on the right side. Write 22 + 13 = and have your child write the sum. Then turn the tray so the 13 objects are on the left side. Have your child look at the new arrangement, and ask him or her to write a different addition equation to represent the addends and find the sum. Finally, ask your child how addend order affects the sum. Repeat using other addends.

Additional Practice

Name _____

Review

You can decompose both addends by place value to find partial sums to help you add two numbers.

34 + **23**	Add the **tens**. $30 + 20 = 50$
$30 + 4 \qquad 20 + 3$	Add the ones. $4 + 3 = 7$
	Add the partial sums. $50 + 7 = 57$
	So, $34 + 23 = 57$.

---

1. How can you decompose both addends by place value? Circle your answer.

   $25 + 13 = ?$

   $25 + 13$ $25 + 13$
   $20 + 3 \quad 10 + 5$ $20 + 5 \quad 10 + 3$

2. Decompose both addends. Then add the partial sums.

   $44 \qquad + \qquad 18 = ?$

   _____ + _____   _____ + _____

   tens: _____ + _____ = _____

   ones: _____ + _____ = _____

   partial sums: _____ + _____ = _____

How can you decompose the addends to find the sum?

3. $26 + 42 =$ _____

4. $39 + 47 =$ _____

5. Explain how to decompose both addends in $65 + 17$.

6. Piper bikes 62 miles in June and 35 miles in July. She says she biked 95 miles in all. How do you respond to Piper? Decompose the addends to explain your thinking.

Math @ Home Activity

Give your child 2 piles of at least 11 pennies each. The total number of pennies should be fewer than 100. (You may also use or cut out different small objects.) Have your child sort the pennies in each pile into groups of ten and some leftover ones to show decomposing both addends. Then encourage your child to find the total number. Repeat the activity several times with different addends.

Additional Practice

Name _____

Review

You can use a number line to add two addends.

$14 + 5 = ?$

One way to add on a number line is to use bars to show the addends.

The sum is the length of both bars put together.

$$14 + 5 = 19$$

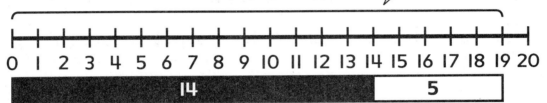

Another way to add on a number line is to use jumps to show the addends.

The number you end on is the sum.

$$14 + 5 = 19$$

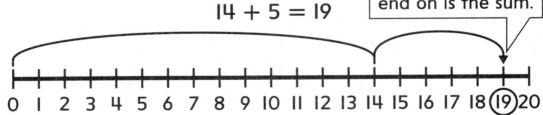

---

1. How can you use the number line to add? Fill in the numbers to complete the equation.

$26 + \underline{\hspace{1cm}} = \underline{\hspace{1cm}}$

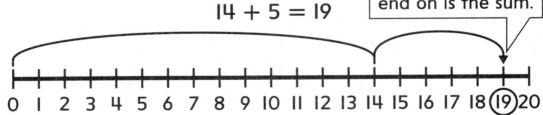

What is the sum? Use the number line to add.

2. 24 + 53 = _____

3. 66 + 19 = _____

---

4. Cassie read 26 pages in her book on Monday and 38 pages on Tuesday. How many pages did she read Monday and Tuesday? Use the number line to help you find the sum.

_____

---

5. How can you use a number line to add two 2-digit numbers?

Additional Practice

Name _____

Review

You can decompose one addend to help you add 2-digit numbers. A number line can help you add.

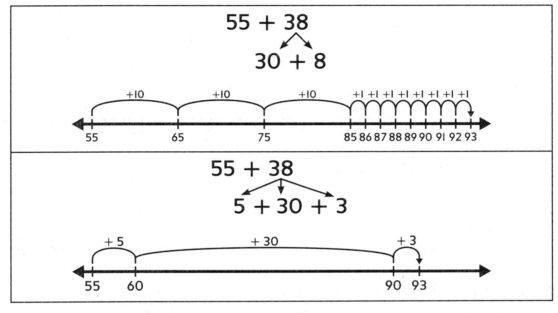

So, 55 + 38 = 93.

How can you decompose one addend to help you add? Circle your answer.

1. 28 + 27 = ?

  20 + 2   20 + 7

2. 46 + 29 = ?

  20 + 8   10 + 10 + 9

3. 54 + 18 = ?

  6 + 10 + 2   50 + 3

4. 65 + 36 = ?

  60 + 3   5 + 30 + 1

5. How can you find the sum by decomposing one addend? Fill in the numbers.

$\underline{44} + 47 = ?$

$44 = 3 + \underline{\hspace{1cm}} + 1$

$47 + 3 + \underline{\hspace{1cm}} + 1 = \underline{\hspace{1cm}}$

Decompose one addend and use the number line to show the addition. Fill in the sum.

6. $33 + 66 = \underline{\hspace{1.5cm}}$

7. $45 + 39 = \underline{\hspace{1.5cm}}$

8. Leland has 28 crayons and 36 markers. He will decompose 28 into $4 + 20 + 4$ to find the total number of crayons and markers. What are two more ways he can decompose one addend to find the total number of crayons and markers?

Have your child decompose one addend while adding two 2-digit numbers. The sum of the numbers should be less than 100. Draw a number line from 0–100. Ask your child to use a number line to find each sum and to explain how he or she found each sum. Repeat several times using different addends.

Additional Practice

Name _____

Review

You can make adding easier by adjusting addends so at least one is a friendly number.

Add an amount to one addend, and then subtract the same amount from the other addend.

$48 + 39 = ?$	$48 + 39 = ?$
[+2] [−2]	[−1] [+1]
$50 + 37 = 87$	$47 + 40 = 87$

---

1. How can you adjust the addends for friendlier addition? Circle all the ways to adjust the addends.

   $27 + 18 = ?$

   $30 + 15$ $30 + 20$ $24 + 20$ $25 + 20$ $30 + 21$

Fill in the boxes to show how the addends were adjusted. Then write the sum.

2.   28   +   33 = ?

   + [_____]   − [_____]

       30   +   31 = _____

3.   47   +   25 = ?

   + [_____]   − [_____]

       50   +   22 = _____

Adjust the addends for friendlier addition. Fill in the adjustments, and complete the equations.

4.   36   +   38 = ?

$-$ ☐
↓
☐   +   ☐   =   ☐

+ ☐
↓

5.   23   +   59 = ?

$-$ ☐
↓
☐   +   ☐   =   ☐

+ ☐
↓

6. How can you adjust the addends to find the sum? Show the addition on the number line.

19 + 68 = _____

7. There are 22 students at a soccer game. Then 49 more students arrive. How can you adjust the addends in the equation 22 + 49 = ? to find the total number of students at the soccer game? Explain.

Math @ Home Activity

With your child, look at the prices of some items advertised on sale. Write an addition equation with a missing sum corresponding to the prices of two different products. Use 2-digit addends with a sum less than 100. Ask your child to explain how to adjust the addends to add more easily and find the sum. Finally, ask him or her to explain another way to adjust the addends to find the same sum.

Additional Practice

Name _____

Review

You can use addition strategies you already know to add more than two 2-digit addends.

$$36 + 33 + 28 = ?$$

One Way You can decompose two addends.

$$36 + 33 \quad + \quad 28 = 97$$

30 + 3   20 + 8

```
        + 30              + 20        + 3  + 8
  |                  |              | |       |
  36                 66            86 89     97
```

Another Way You can adjust the addends.

$$33 + 28 + 36 = 97$$ You can change the order of addends to solve.

$33 + 28 = ?$	$61 + 36 = ?$
−2 +2	−1 +1
31 + 30 = 61	60 + 37 = 97

So, $36 + 33 + 28 = 97$

---

1. Juan is adding $47 + 28 + 13$. How can you change the order of the addends to make friendly numbers?

   47 + _____ + _____

What is the sum? Use an addition strategy to solve.

2. $22 + 25 + 38 =$ _____

3. $11 + 23 + 38 + 17 =$ _____

---

4. Last week, Lena practiced the piano for 17 minutes on Monday, 32 minutes on Wednesday, 23 minutes on Thursday, and 19 minutes on Friday. How many minutes did Lena practice the piano last week?

   _____ minutes

5. Grace has 18 yellow, 36 purple, 29 blue, and 12 red beads. How many beads does she have in all? Explain how you found the sum.

   _____ beads

Math @ Home Activity

Have your child practice adding more than two 2-digit numbers by adding the number of minutes he or she spends on a certain activity in a week. The sum should be within 100. For example, your child can add the minutes spent playing a sport on three different days. Then encourage your child to explain his or her addition strategy.

Additional Practice

Name _____

Review

You can use addition strategies to solve one- and two-step word problems.

Ian has $25. He gets $18 for his birthday and earns $14 painting. How much money does Ian have now?

You can use base-ten shorthand and an equation to represent the problem.

$25 + $18 + $14 = ?

One way to solve is to decompose. $25 + $18 + $14 = $57

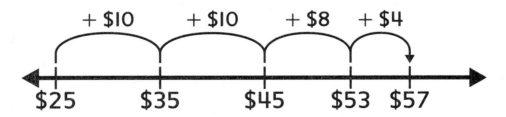

1. Represent the problem using base-ten shorthand. Complete the equation.

Tina has 19 pears, 27 apples, and 41 oranges to make fruit baskets. How many pieces of fruit does Tina have? _____ + 27 + _____ = _____

2. There are 38 students in the gym. There are 21 more students in the library than in the gym. How many students are in the library?

_____

3. Jason has 23 baseball cards. His brother has 12 more baseball cards than Jason. How many baseball cards do Jason and his brother have in all? Choose a strategy to solve the problem. Explain or show how you used the strategy.

_____

Math @ Home Activity

Look for opportunities around your home where your child can practice solving two-step addition problems. For example, ask your child to find the total value of his or her allowance for 3 months or have your child find the total amount of time he or she spends practicing a sport or instrument in 4 weeks.

Additional Practice

Name _____

Review

You can count on or count back to subtract.

15 − 7 = ?

Count on to subtract.

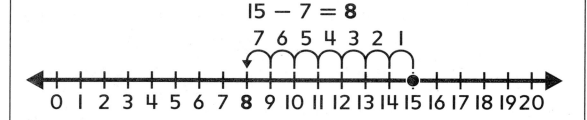

15 − 7 = **8**

Count back to subtract.

15 − 7 = **8**

---

1. How can you use the number line to count on to subtract? Fill in the difference.

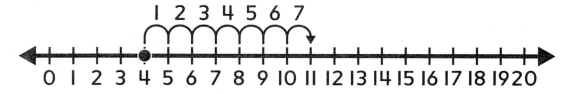

11 − 4 = _____

2. How can you use the number line to count back to subtract? Fill in the difference.

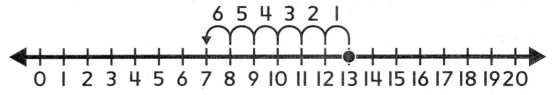

$13 - 6 =$ _____

3. How can you use the number line to find the difference? Fill in the difference.

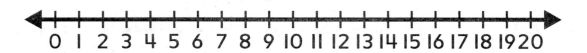

$17 - 9 =$ _____

4. Write a word problem that can be represented by the number line. Write an equation that represents the problem. Then solve.

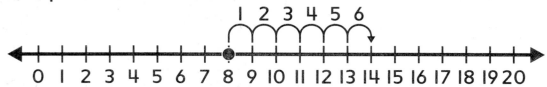

Math @ Home Activity

Draw a number line that goes from 0 to 20, like the number line in Exercise 3. Pose a subtraction word problem, like the following, to your child. If your child has 12 math problems for homework and completes 5 of the problems, how many problems are left? Ask your child to use the number line to find the difference. Have your child explain how the number line was used to solve the problem.

Lesson 6-2

Additional Practice

Name _____

Review

You can make a 10 or use addition to fluently subtract within 20.

There are 13 pages in a chapter. Aidan reads 7 pages. How many pages are left to read in the chapter?

Decompose the number being subtracted to make a 10.	$13 - 7 = 6$ $3 + 4$

Use addition to subtract.

$7 + 6 = 13$

1 2 3 4 5 6

0 1 2 3 4 5 6 7 8 9 10 11 12 13 14 15 16 17 18 19 20

$13 - 7 = 6$

There are 6 pages left to read in the chapter.

---

1. How can you make a 10 to subtract? Fill in the numbers.

$17 - 9 = ?$

___ ___

$17 - 9 = ?$

$17 - $ ___ $ = 10$

$10 - $ ___ $ = 8$

$17 - 9 = $ ___

How can you rewrite the equation as an addition equation with an unknown addend? Fill in the numbers.

2. $11 - 5 = ?$

   _____ $+ ? = 11$

   $11 - 5 =$ _____

3. $16 - 8 = ?$

   _____ $+ ? = 16$

   $16 - 8 =$ _____

---

4. Brian is doing a set of 15 squats. He has done 8 squats. How many squats does he have left in the set? Show your work using addition to subtract.

5. Lucy has 13 apple slices. She gives her sister 7 apple slices. How can she make a 10 to find the number of apple slices she has left?

Math @ Home Activity

Have your child write a subtraction word problem about dinner or a snack. For example, your child has 17 peas on his or her plate. After your child eats 9 peas, how many are left? Tell your child to use one of the strategies that was taught in this lesson to solve the problem. Ask your child to explain how he or she found the answer.

Lesson 6-3

Additional Practice

Name _____

Review

You can use base-ten blocks to represent and solve 2-digit subtraction equations.

Ricky buys 24 pencils. He gives 11 pencils to his friends. How many pencils does he have left?

Represent 24 with base-ten blocks.

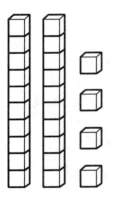

Then take away 1 ten and 1 one.

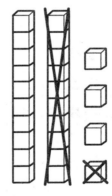

So, 24 − 11 = 13.

Ricky has 13 pencils left.

What is the difference? Cross out base-ten blocks. Then fill in the difference.

1. 38 − 15 = _____

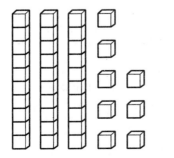

2. _____ = 55 − 23

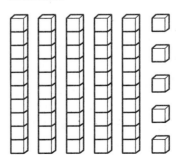

What is the difference? Draw base-ten shorthand. Then fill in the difference.

3. $47 - 25 = $ _____

4. _____ $= 59 - 36$

---

5. Christy is watching a movie that is 96 minutes long. She has watched 43 minutes of the movie. How many minutes of the movie are left?

6. Riley has 65 inches of ribbon. She uses 32 inches of ribbon to decorate a page in her scrapbook. She draws these blocks to find how many inches of ribbon she has left. How can you help Riley fix her drawing?

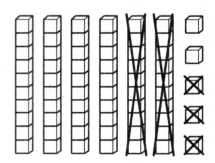

Math @ Home Activity

Give your child 100 pennies. Have your child use the pennies to subtract 2-digit numbers. To deepen your child's understanding, subtract incorrectly and have your child show you how to correct the mistake.

Additional Practice

Name _____

Review

You can use base-ten blocks to represent and solve 2-digit subtraction equations with regrouping.

Represent and solve 45 − 17.

Represent 45 with base-ten blocks.	Regroup 1 ten into 10 ones. Take away 1 ten and 7 ones.
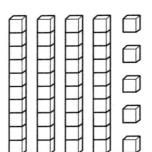	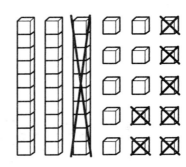

So, 45 − 17 = **28**.

Is regrouping needed to subtract? Circle Yes or No.

1. 32 − 19

    Yes   No

2. 58 − 21

    Yes   No

3. 65 − 43

    Yes   No

4. 77 − 38

    Yes   No

What is the difference? Draw base-ten shorthand. Then fill in the difference.

5. 31 − 12 = _____

6. 54 − 28 = _____

7. Borut is 62 inches tall. His brother is 39 inches tall. How much taller is Borut than his brother?

8. Write a 2-digit subtraction equation that requires regrouping to solve. How do you know regrouping is needed to solve the equation you wrote?

Math @ Home Activity

Look in a catalog or online for the price of an item your child would like to buy. The price should be a 2-digit number. Give your child a 2-digit money amount he or she can spend that would require regrouping to find the amount that would be left over. Have your child draw base-ten shorthand to find the amount of money that would be left over after buying the item.

Additional Practice

Name _____

Review

You can use a number line to subtract.

Subtract 16 − 9.

One way to subtract is to use bars on a number line.

The difference is what remains from the longer bar.

$16 - 9 = 7$

16

| | 9 |

0 1 2 3 4 5 6 7 8 9 10 11 12 13 14 15 16 17 18 19 20

Another way to subtract is to make jumps on a number line.

The number you end on is the difference.

$16 - 9 = 7$

0 1 2 3 4 5 6 7 8 9 10 11 12 13 14 15 16 17 18 19 20

I. How can you use a number line to subtract? Fill in the difference.

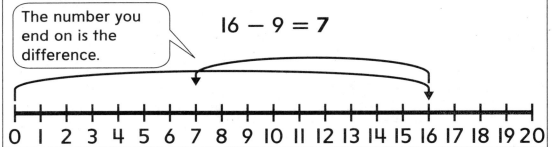

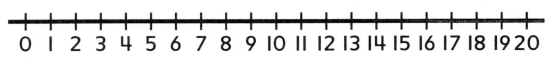

$11 - 5 =$ _____

What equation matches the subtraction shown on the number line? Fill in the equation.

2. ____ − ____ = ____

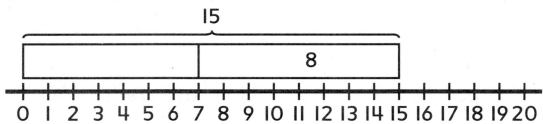

15

| | 8 |

0 1 2 3 4 5 6 7 8 9 10 11 12 13 14 15 16 17 18 19 20

3. ____ − ____ = ____

0 1 2 3 4 5 6 7 8 9 10 11 12 13 14 15 16 17 18 19 20

What is the difference? Use the number line.

0 5 10 15 20 25 30 35 40 45 50 55 60 65 70 75 80 85 90 95 100

4. 58 − 37 = ____   **5.** 84 − 56 = ____

6. Complete the equation and explain how you can use a number line to subtract.

____ − 45 = 22

Math @ Home Activity

Make a number line from 0 to 20 by drawing on your sidewalk or driveway with chalk or by using painter's tape on a floor inside your home. Have your child find 14 − 6 by making jumps on the number line. Tell your child to start on 14 and then make a jump of 6 to the left. Repeat the activity, this time having your child find 17 − 8.

Student Practice Book

74

Additional Practice

Name _____

Review

You can decompose a number by place-value to help you solve a 2-digit subtraction equation.

Olga has 43 apples and sells 25 of them. How many does she have left?

You can subtract $43 - 25$.

Decompose 25 by place value.	**43 − 25**

$$20 + 5$$

Use an open number line to count back the tens and ones.

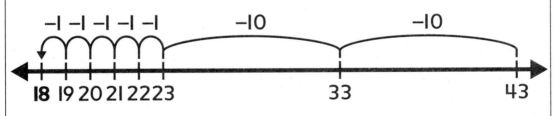

So, $43 - 25 = 18$.

Is the number decomposed by place value?
Circle Yes or No.

1. $62 - 39 = ?$

$$3 + 9$$

Yes   No

2. $84 - 57 = ?$

$$50 + 7$$

Yes   No

How can you decompose by place value to find the difference? Show the subtraction on the number line.

3. 51 − 16 = ___

___ + ___

4. 75 − 28 = ___

___ + ___

---

5. Can you subtract 56 − 28 by decomposing 28 into 2 and 8? Explain your thinking.

6. Lola's hair is 32 inches long. She is getting 14 inches cut off. Write step-by-step instructions explaining how long her hair will be after she gets it cut. Decompose a number to find the difference. Fill in the difference.

32 − 14 = _____

Additional Practice

Name _____

Review

You can adjust numbers to make them friendlier to subtract.

Gabe makes a paper chain with 63 red links and 38 blue links. How can you find how many more red links there are than blue links?

You can subtract 63 − 38 to compare.

One way:	Another way:
Subtract 3 to adjust the numbers.	Add 2 to adjust the numbers.
63 − 38 = 25	63 − 38 = 25
−3 −3	+2 +2
60 − 35 = 25	65 − 40 = 25

Adjust both numbers the same way.

Solve the friendlier equation to find the difference.

There are 25 more red links than blue links.

How can you adjust the numbers for friendlier subtraction? Complete the equations.

1. 57 − 32 = ___

    + ___   + ___

    ___ − ___ = ___

2. 81 − 46 = ___

    − ___   − ___

    ___ − ___ = ___

3. Which ways show how to adjust the numbers to subtract? Choose all the correct answers.

$$43 - 28$$

A. $40 - 25$ **B.** $40 - 31$

C. $45 - 30$ **D.** $41 - 30$

---

4. Show two ways to adjust the numbers in $92 - 69$. Then find the difference.

5. Ona has $77 and spends $51 on clothes. She finds how much money she has left by adjusting the numbers and finding the difference of $76 − $50. She says she has $26 left. Is Ona adjusting the numbers correctly? Explain your thinking.

Math @ Home Activity

Have your child solve $42 - 28 = ?$ by adjusting the numbers. Before moving on to the next problem, have your child explain how your child could have adjusted the numbers in a different way. Repeat the activity with $67 - 33 = ?$.

Additional Practice

Name _____

Review

You can find the difference of a subtraction equation by writing and solving a related addition equation.

There are 64 students in second grade at Greenwood Elementary School. 29 of the students are boys. How can you find the number of girls?	Whole	
	64	

Whole	
64	
Part	Part
29	?

Write a subtraction equation with the difference unknown.	Write a related addition equation with an unknown addend.
$64 - 29 = ?$	$29 + ? = 64$

The unknown will be the same for both related equations. The difference and the unknown addend are both 35.

How can you use the Part-Part-Whole mat to fill in the subtraction equation and the related addition equation?

1.

Whole	
47	
Part	Part
18	?

$47 - \underline{\quad} = ?$

$? + 18 = \underline{\quad}$

2.

Whole	
83	
Part	Part
?	36

$\underline{\quad} - 36 = ?$

$? + \underline{\quad} = \underline{\quad}$

3. What related addition equation can you use to find the difference? Choose the correct answer.

$$52 - 27 = ?$$

A. $52 + 27 = ?$ **B.** $27 + 52 = ?$

C. $52 - ? = 27$ **D.** $27 + ? = 52$

---

4. A flower shop has 96 roses. The shop sells 68 roses. How many roses are left? Fill in the numbers.

___ − ___ = ?

___ + ? = ___

There are ___ roses left.

5. Emerson has a soccer game on Saturday that lasts 45 minutes. She rides her bike on Sunday. She exercises for a total of 71 minutes on Saturday and Sunday. How many minutes does she ride her bike on Sunday? Use an addition equation to solve the subtraction problem. Explain your thinking.

Math @ Home Activity

Pose an everyday problem to your child that involves subtracting 2-digit numbers. Have your child use an addition equation to solve the problem. Then ask your child to explain how the addition equation relates to the problem.

Lesson 6-9
Additional Practice

Name _____

Review

You can use subtraction strategies to solve one-step word problems.

Gwen is going to visit her grandparents. They live 67 miles away. Gwen's family has driven 39 miles. How can you find how many miles are left?

First, make sense of the problem and represent it with an equation.	Then, use a subtraction strategy to solve it.
⊢-------67 miles--------⊣ **Total miles** **Miles driven** | **Miles left** ⊢----39 miles----⊢---? miles--⊣ $67 - 39 = ?$	$67 - 39 = 28$ You can adjust the numbers for friendlier subtraction. $\boxed{+1}$ $\boxed{+1}$ $68 - 40 = 28$

There are 28 miles left.

---

1. What equation represents the problem? Choose the correct answer.

Eli's black dog weighs 72 pounds and his brown dog weighs 56 pounds. Which can be used to find how many more pounds the black dog weighs than the brown dog?

A. $56 - 72 = ?$ **B.** $72 + 56 = ?$

C. $72 + 16 = ?$ **D.** $72 - 56 = ?$

How can you represent and solve the word problem? Fill in the equation and use any strategy to solve.

2. Jamal has 48 crayons. He gives his brother 23 crayons. How many crayons does Jamal have left?

_____ − _____ = _____

3. A play is 95 minutes long. Amal has watched 66 minutes of the play. How many minutes are left?

_____ − _____ = _____

---

4. Pete is finding 51 − 17. His work is shown. How would you respond to Pete?

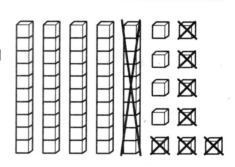

Have your child be the teacher and create a one-step subtraction word problem that involves 2-digit numbers. As the teacher, your child should explain to you how to solve the problem by using a strategy that was taught in this unit.

Additional Practice

Name _____

Review

You can use subtraction strategies to solve two-step word problems.

Yi buys 31 pieces of fruit. He buys 13 apples, 9 bananas, and the rest are oranges. How many oranges does Yi buy?

Use a subtraction strategy to solve it.

$31 - 13 - 9 = ?$

10 + 3

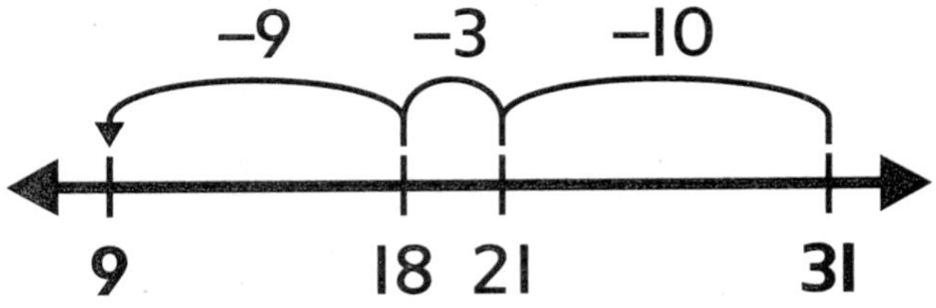

Yi buys 9 oranges.

I. How can you represent and solve the word problem? Use any strategy to solve.

Erin sells 44 books. She sells 26 mystery books, 8 fantasy books, and some comic books. How many comic books does she sell?

How can you represent and solve the word problem? Use any strategy to solve.

2. Darron has $62. He buys pants for $29 and a shirt for $14. How much money does Darron have left?

3. Ava will bike a total of 87 miles over three days. On the first day she will bike 35 miles. She will bike 21 miles on the second day. How many miles will Ava bike on the third day?

4. a. Write a two-step subtraction word problem.

   b. Use any strategy to solve your problem.

Math @ Home Activity

Look for situations around your home where your child can solve two-step word problems with subtraction. Have your child show you how to solve the problems using strategies from this unit. For example, if your child needs to read a total of 78 pages over three days, and 21 pages were read one day and 40 pages the next, how many pages need to be read on the third day?

Additional Practice

Name _____

Review

You can use an inch ruler to measure the length of an object. Inches is the unit of measure.

What is the length of the glue stick?

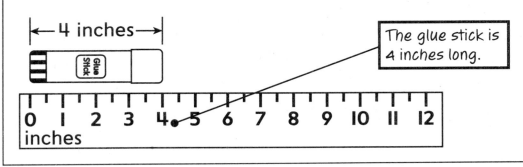

The glue stick is 4 inches long.

What is the length of the object? Use an inch ruler to measure.

1.

___ ___

2.

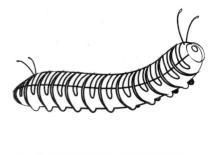

___ ___

---

3. Will this spoon fit into a box that has a length of 2 inches? Explain your thinking.

4. Althea says that her key is 8 inches long. How do you respond to Althea?

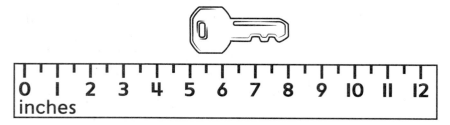

5. How do you measure the length of a paper clip using an inch ruler?

Math @ Home Activity

Have your child measure the lengths of household objects to the nearest inch. For example, before putting away groceries, your child can measure the lengths of boxed foods.

Additional Practice

Name _____

Review

You can use different tools to measure length.

An inch ruler measures length in inches or feet.

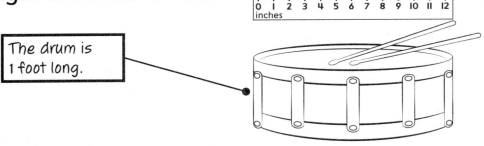

The drum is 1 foot long.

A yardstick and a measuring tape measure length in inches, feet, or yards.

The lizard is 2 feet long.

1. What is the length of the hammer?

____ _____

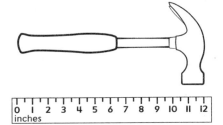

2. What is the length of the bat?

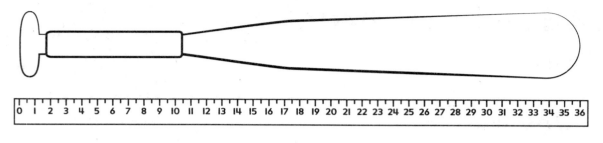

0 1 2 3 4 5 6 7 8 9 10 11 12 13 14 15 16 17 18 19 20 21 22 23 24 25 26 27 28 29 30 31 32 33 34 35 36

_____ _____

_____

Which is the best tool to use for the measurement? Circle your answer.

3. length of a cell phone

    A. ruler   **B.** yardstick   **C.** tape measure

4. length of a football field

    A. ruler   **B.** yardstick   **C.** tape measure

5. What unit would you use to measure the length of a doll? Explain.

6. What tool would you use to measure the length of a cafeteria table? Explain your thinking.

Math @ Home Activity

Look for household items that can be easily measured with a ruler or yardstick. Ask your child to determine the appropriate measuring tool and then find the actual measurement. Encourage your child to record their findings.

Additional Practice

Name _____

Review

You can compare lengths by subtracting the measurements to find the difference.

Measure each pencil and write a subtraction equation to compare them.

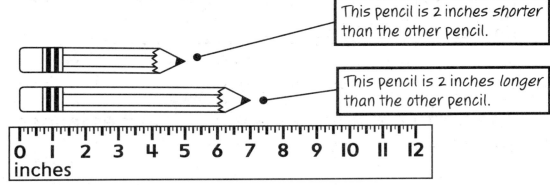

This pencil is 2 inches *shorter* than the other pencil.

This pencil is 2 inches *longer* than the other pencil.

$7 - 5 = 2$

How can you compare the lengths? Fill in the equation.

1. A slug moves 4 inches. A snail moves 8 inches.

   _____ – _____ = _____

2. Jamal's paper airplane flew 6 yards. Camille's paper airplane flew 9 yards.

   _____ – _____ = _____

3. Hank's table is 5 feet long. Darcy's table is 3 feet long.

   _____ – _____ = _____

89

Copyright © McGraw-Hill Education

4. Which object is longer? Fill in the equation and the answer.

_____ − _____ = _____

The bottom sunglasses are _____ inches longer than the top sunglasses.

---

5. Haddy can throw a baseball 5 yards. Liam can throw a baseball 7 yards. How much farther can Liam throw a baseball than Haddy? Explain.

Math @ Home Activity

Have your child compare lengths of objects in inches, feet, or yards. For example, ask your child to determine how many feet longer the kitchen table is than the coffee table.

Additional Practice

Name _____

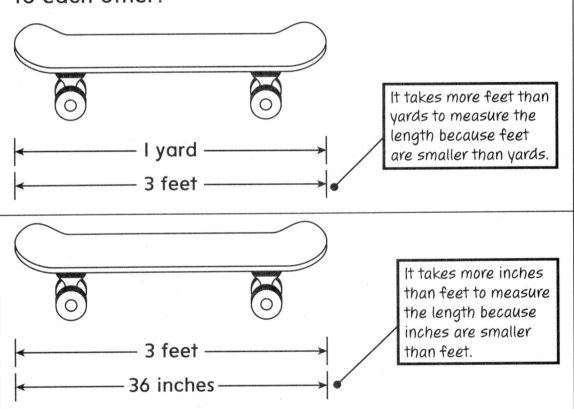

Review

You can measure length in inches, feet, or yards. The smaller the unit, the more units are needed to measure an object's length.

How do the measurements of the skateboard relate to each other?

├─────── 1 yard ───────→
├─────── 3 feet ───────→

It takes more feet than yards to measure the length because feet are smaller than yards.

├─────── 3 feet ───────→
├─────── 36 inches ──────→

It takes more inches than feet to measure the length because inches are smaller than feet.

1. What is the length of a window in inches?

_____ inches

Will the measurement of a window have more inches or yards? Circle the correct answer.

inches   yards

2. What is the length of a door in yards?

_____ yards

Will the measurement of a door have fewer feet or fewer yards? Circle the correct answer.

feet   yards

---

3. Carol measured the length of her bed twice using inches and feet. Will Carol's measurements have more inches or feet? Explain.

4. The length of a snake at a zoo is being measured in feet and then in yards. Will there be more feet or yards in the measurements? Explain your reasoning.

Math @ Home Activity

Look for opportunities for your child to relate inches, feet, and yards at home. For example, ask your child to measure the length of a table in inches and then measure that length in feet.

Additional Practice

Name _____

Review

You can use everyday objects to help you estimate length in inches and feet.

A paper clip is about I inch.	A math book is about I foot.

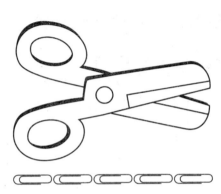

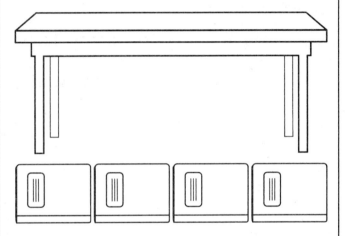

So, the scissors are about 5 inches long.

So, the table is about 4 feet long.

Which everyday item can you use to estimate the length of the object? Circle the correct answer.

I. couch

   paper clip

   math book

2. cereal box

   paper clip

   math book

3. index card

   color tile

   science book

4. hockey stick

   color tile

   science book

5. How long is the shoe? Estimate the length.

_____ _____

6. How long is the jump rope? Estimate the length.

_____ _____

---

7. Evan uses a clipboard to estimate the length of his bedroom. He says his room is about 12 inches long. Is Evan's estimate reasonable? Explain your thinking.

Math @ Home Activity

Have your child estimate the lengths of three toys to the nearest inch. Make sure the toys vary in length. Your child can line up paper clips or quarters end to end to estimate the lengths of the toys.

Additional Practice

Name _____

Review

You can use different tools to measure length in metric units.

A centimeter ruler measures length in centimeters.

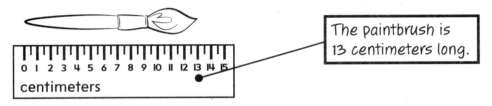

The paintbrush is 13 centimeters long.

0 1 2 3 4 5 6 7 8 9 10 11 12 13 14 15
centimeters

A meterstick measures length in meters.

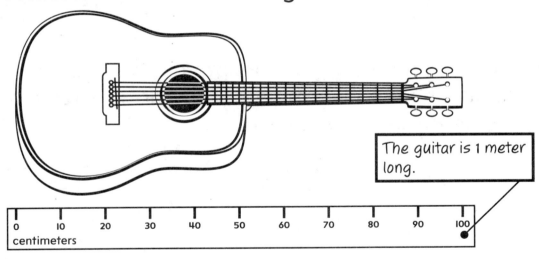

The guitar is 1 meter long.

0   10   20   30   40   50   60   70   80   90   100
centimeters

What is the length of the object in centimeters? Use a centimeter ruler to measure.

1.

2.

_____ _____

What is the length of the object in meters? Use a meterstick to measure.

3. window

_____ _____

4. door

_____ _____

5. couch

_____ _____

---

6. Use a centimeter ruler to draw a marker that is 14 centimeters long.

7. Carlos is measuring the length of his swing set. Should he use a centimeter ruler or a meterstick to measure? Explain why.

Math @Home Activity

Explain to your child that a fingernail is about 1 centimeter wide and 1 meter is about the length of your child's outstretched arms. Have your child identify objects around the house that can be measured in centimeters and in meters. Then have your child estimate and measure one object in centimeters and one object in meters.

Additional Practice

Name _____

Review

You can compare lengths by subtracting the measurements to find the difference.

Debra wants to use the shorter piece of yarn for a craft. What is the difference in the lengths of the yarn pieces?

You can measure each string and write a subtraction equation to compare them.

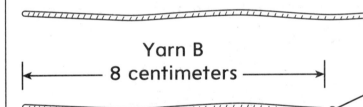

Yarn A is 6 centimeters longer than Yarn B.

Yarn A
— 14 centimeters —

Yarn B is 6 centimeters shorter than Yarn A.

Yarn B
— 8 centimeters —

$14 - 8 = 6$

How can you compare the lengths? Fill in the equation.

1. Kai's pencil box is 22 centimeters long. His pencil is 16 centimeters long.

   _____ − _____ = _____

2. The classroom is 12 meters long. The gymnasium is 35 meters long.

   _____ − _____ = _____

3. How much longer is the bottom leaf than the top leaf?

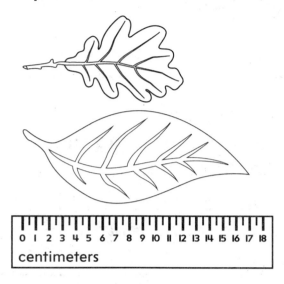

centimeters

_____ _____

---

4. The length of a dump truck is 7 meters. A grader is 15 meters long. How much shorter is the dump truck than the grader?

5. Emily's pink hair ribbon is 43 centimeters long. Her red hair ribbon is 27 centimeters long. How much longer is the pink hair ribbon than the red hair ribbon? Explain your thinking.

Math @ Home Activity

Provide your child with different opportunities to measure objects in centimeters and meters. While visiting a playground, encourage your child to measure plant life in centimeters and structures in meters. Then have your child determine how much longer one object is than the other object.

Additional Practice

Name _____

Review

You can measure length in centimeters and meters. The smaller the unit, the more units are needed to measure an object's length.

Juan measured the table in meters and centimeters.

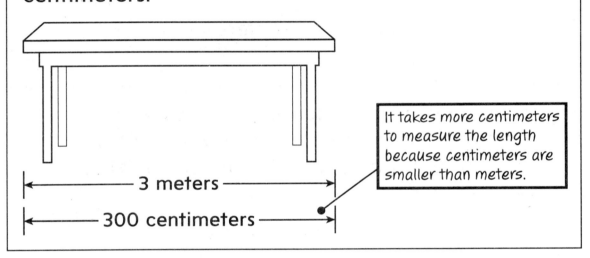

It takes more centimeters to measure the length because centimeters are smaller than meters.

├────── 3 meters ──────┤

├────── 300 centimeters ──────┤

---

1. What is the length of a wall in meters?

   _____ meters

   Will the measurement of the wall have more centimeters or more meters? Circle the correct answer.

   centimeters   meters

2. What is the length of a bookshelf in centimeters?

_____ centimeters

Will the measurement of the bookshelf have fewer centimeters or fewer meters? Circle the correct answer.

centimeters   meters

3. The length of a dresser is measured in meters and centimeters. Will the measurements have more meters or more centimeters? Circle the correct answer.

centimeters   meters

---

4. Niles measured the length of a poster in centimeters. Then he measured it in meters. Are there more centimeters or meters? Explain your thinking.

Math @ Home Activity

Have your child measure the length of a couch in centimeters and meters. Ask your child to explain why there are more centimeters than meters in the measurements.

Additional Practice

Name _____

Review

You can use everyday objects to help you estimate length in centimeters and meters.

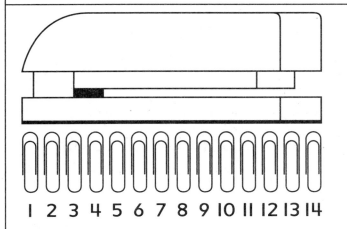

The end of a paper clip is about 1 centimeter.

So, the stapler is about 14 centimeters long.

1 2 3 4 5 6 7 8 9 10 11 12 13 14

The length of a second grader's wrist to their other wrist is about 1 meter.

So, the living room is about 5 meters long.

Which unit would you use to measure the length of the object? Circle the correct answer.

1. playground

    centimeter   meter

2. calculator

    centimeter   meter

Which everyday item can you use to estimate the length of the object? Circle the correct answer.

3. trampoline

    unit cube

    baseball bat

4. eraser

    width of paperclip

    arm span

5. worm

    staple

    baseball bat

6. hose

    unit cube

    arm span

---

7. Kayla estimates that her rabbit is 22 meters long. Is Kayla's estimate reasonable? Explain your thinking.

8. Ray wants to estimate the length of a baseball card. Should Ray use the width of a paper clip or his arm span to find the estimate? Explain.

Math @ Home Activity

Explain to your child that a fingernail is about 1 centimeter wide and 1 meter is about the length of your child's outstretched arms. Have your child estimate the lengths of household objects in centimeters and meters. Ask your child to explain why he or she chose centimeters or meters when estimating the length of each object.

Additional Practice

Name _____

Review

You can solve addition and subtraction word problems involving length.

Anna has 11 inches of string. After she cuts a piece of string to make a bracelet, there are 4 inches of string left. How long is the piece of string Anna cut?

You can make a drawing to represent the problem.

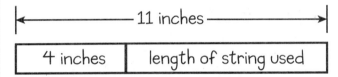

Write an equation. $11 - 4 = ?$

Solve the equation. $11 - 4 = 7$

The piece of string Anna cut is 7 inches long.

I. What operation would you use to solve the problem? Circle the correct answer.

Abby is riding on a zip line. The zip line is 17 meters long. She has gone 8 meters. How many meters does she have left to go?

addition   subtraction

How can you represent the problem with a drawing and an equation? Solve the equation.

2. Camden's train track is 46 inches long. His brother's train track is 39 inches long. How long are the boys' train tracks in all?

3. Everett's pet lizard is 13 centimeters long. Mark's pet lizard is longer. The difference in length between the lizards is 5 centimeters. How long is Mark's lizard?

4. Vera is making a keychain. She has a cord that is 34 inches long. She cuts 16 inches off the cord that she does not use. Explain how you would find how much cord Vera used.

Math @ Home Activity

Provide opportunities for your child to solve addition and subtraction word problems involving lengths of household objects. For example, have your child choose two objects. Write a word problem for your child to solve that involves adding or subtracting the lengths of the objects. Have your child draw a picture to represent the problem. Then tell your child to write and solve an equation to solve the problem. Repeat several times.

Additional Practice

Name _____

Review

You can solve addition and subtraction word problems involving length.

A frog hops 14 inches. A toad hops 32 inches. How many more inches does the toad hop than the frog?

You can make a drawing to represent the problem.

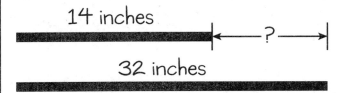

14 inches

?

32 inches

Write an equation. $32 - 14 = ?$

Solve the equation. $32 - 14 = 18$

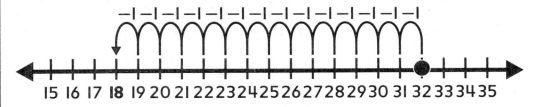

15 16 17 **18** 19 20 21 22 23 24 25 26 27 28 29 30 31 32 33 34 35

The toad hops 18 inches more than the frog.

---

1. The length of Olivia's skateboard is 71 centimeters. The length of one of Olivia's roller skates is 16 centimeters. How much longer is the skateboard? Circle the equation you can use to solve the problem.

$$71 + 16 = ? \qquad 71 - 16 = ?$$

2. How can you make a drawing and write an equation to represent the problem? Use the number line to solve.

Addy's paper chain is 29 inches long. Mia's paper chain is 12 inches shorter than Addy's paper chain. How long is Mia's paper chain?

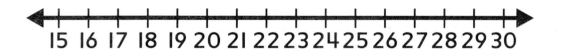

15 16 17 18 19 20 21 22 23 24 25 26 27 28 29 30

3. Lin connects two hoses so she can water her garden. One hose is 32 yards long. The other hose is 25 yards long. How many yards long are the two hoses in all? Explain your thinking.

Math @ Home Activity

Have your child write an addition or subtraction word problem that involves the lengths of two of his or her toys. Ask your child to draw a picture and write an equation to represent the problem. Then tell your child to solve the equation.

Additional Practice

Name _____

Review

You can skip count to find the total value of a group of the same type of coin.

Carmen has 40 cents in her piggy bank.
What coins might she have?

 40 pennies = 40¢

         8 nickels = 40¢

5¢ 10¢ 15¢ 20¢ 25¢ 30¢ 35¢ 40¢

     4 dimes = 40¢

10¢ 20¢ 30¢ 40¢

So, Carmen might have 40 pennies, 8 nickels, or
4 dimes in her piggy bank.

Fill in the name of the coin. Then draw a line to match the coin with its value.

1.  _____   1¢

2. _____   5¢

3. _____   10¢

4. _____   25¢

Draw a line to match the group of coins to its value.

5.   4¢

6.   30¢

7.   50¢

8.   60¢

9. Tom has 80¢. All his coins are the same. What coins might Tom have? List all the types possible.

10. Amber has 10 nickels. Jasmine has 2 quarters. Who has more money? Explain your thinking.

Additional Practice

Name _____

Review

You can find the total value of a set of mixed coins by skip counting like coins and then adding the values.

Milford has I quarter and 2 dimes in his pocket. He has 2 quarters, I dime, 2 nickels, and 3 pennies in his piggy bank. How much money does he have in each?

Arrange the coins in order of value. Then skip count the value of each coin.

25,   35,   **45**

Milford has 45¢ in his pocket.

25,   50,   60, 65, 70, 7I, 72, **73**

Milford has 73¢ in his piggy bank.

What is the value of the group of coins?

I.   _____

2.   _____

3. Zina gives 76 cents to her brother. What combination of coins could her brother have?

4. Seema finds 91 cents in her couch. What combination of coins could she find?

5. Julia finds 2 dimes, 1 nickel, and 4 pennies on the street. What is the value of the coins?

6. Berna buys a bottle of water with some of her own money and 2 quarters, 2 nickels, and 1 penny her dad gives her. The bottle of water costs 99¢. How much of her own money does Berna spend? Explain your thinking.

Math @ Home Activity

Give your child a collection of coins. Have him or her pick 6 coins without looking. Then have your child find the value of the 6 coins. Repeat the activity, having your child pick a different number of coins.

Additional Practice

Name _____

Review

You can find the total value of bills or coins by skip counting and then adding the values.

Cristal has $36. She gets $27 for her birthday. What are some ways to show the money she has?

You can use dollar bills. Each bill has a different value.

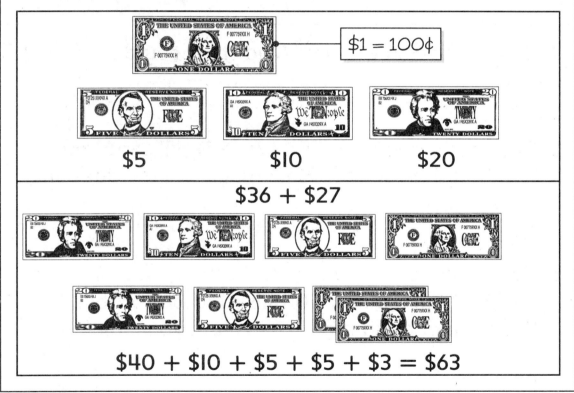

$5 $10 $20

$36 + $27

$40 + $10 + $5 + $5 + $3 = $63

I. What is the value of the group of coins?

_____

What is the value of the group of bills?

2.

_____

3.

_____

4. Devan has $66 dollars. What bills could he have?

5. Gina has three $20 bills, two $10 bills, and four $1 bills in her pocket. How much money does she have in all?

6. Kai wants to buy sneakers that cost $71. She has two $20 bills, three $10 bills, and one $5 bill. Does Kai have enough money? Explain your thinking.

Math @ Home Activity

Give your child a collection of dollar bills to use for this activity. Next, show him or her a picture of an item in a store ad. Tell your child the cost of the item after it is rounded to the nearest whole dollar amount. Then ask your child if there is enough money to buy the item. If yes, have your child find the change left. If no, have your child find how much more money is needed to buy the item.

Additional Practice

Name

Review

You can tell time to the nearest five minutes on an analog clock by skip counting by 5s.

Carly has a dance lesson at 3:45. Then she eats a banana at 4:15. She reads a book at 4:30. The clocks show the times Carly does each activity.

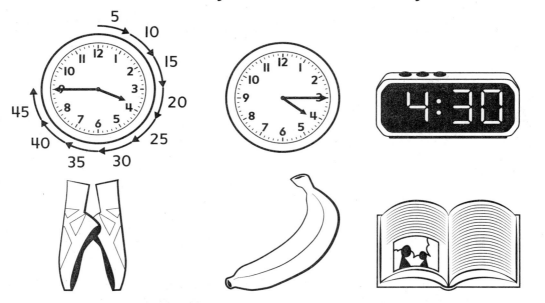

What time is shown on the analog clock? Circle the digital clock that matches.

What time is shown on the analog clock? Write the time.

3.

_____ : _____

4.

_____ : _____

5. What time is shown on the analog clock? Choose all the correct answers.

A. quarter past 1

B. quarter to 1

C.

D.

Math @ Home Activity

Provide opportunities for your child to practice telling time to the nearest 5 minutes at home. While performing everyday activities, ask your child to identify the time when the activity starts and when it ends.

Additional Practice

Name _____

Review

You can use a time line to show the order of events during the day.

The time between midnight and noon is represented using a.m.

The time from noon to midnight is represented using p.m.

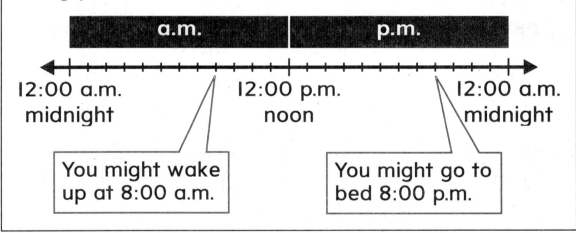

12:00 a.m.
midnight

12:00 p.m.
noon

12:00 a.m.
midnight

You might wake up at 8:00 a.m.

You might go to bed 8:00 p.m.

What time of day does the event take place?
Write *a.m.* or *p.m.*

1.

2.

2:00 _____

4:00 _____

What time of day does the event take place?
Write *a.m.* or *p.m.*

3.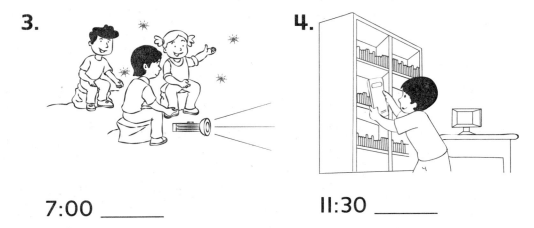

7:00 _____

4.

11:30 _____

---

5. Sook is getting dressed for school. Would the time be 8:15 a.m. or 8:15 p.m.? Explain your thinking.

6. Val tells her mom she has softball practice at 3:00 a.m. How do you respond to Val?

Math @ Home Activity

Have your child identify 6 activities he or she does throughout the day. Write these activities down the left side of a blank sheet of paper. On the right side of the paper, write starting times, with a.m. or p.m., that correspond to the activities. Do not write the correct time for each activity directly across from it. Have your child draw a line from each activity to the correct starting time.

Additional Practice

Name _____

Review

You can use patterns to add 10 and 100 to 3-digit numbers.

How can you help Jordan complete the table?

$224 + 10 = ?$	$224 + 100 = ?$
$234 + 10 = ?$	$234 + 100 = ?$
$244 + 10 = ?$	$244 + 100 = ?$
$294 + 10 = ?$	$294 + 100 = ?$
$394 + 10 = ?$	$394 + 100 = ?$

Adding 10 makes the tens digit go up by 1.

$224 + 10 = 234$
$234 + 10 = 244$
$244 + 10 = 254$
$294 + 10 = 304$
$394 + 10 = 404$

Adding 100 makes the hundreds digit go up by 1.

$224 + 100 = 324$
$234 + 100 = 334$
$244 + 100 = 344$
$294 + 100 = 394$
$394 + 100 = 494$

Is the statement true or false? Explain your answer.

1. Addition patterns can help you add 10 or 100 to a 3-digit number.

2. The ones digit does not change when you add 10 or 100 to a 3-digit number.

What is the sum? Use a number line to show your thinking.

3. $563 + 10 =$ _____

4. $791 + 10 =$ _____

What is the sum?

5. $330 + 10 =$ _____

6. $499 + 10 =$ _____

7. $616 + 100 =$ _____

8. $824 + 100 =$ _____

9. There are 748 people at a museum. Then 10 children and 100 adults enter the museum. How many people are at the museum now? Explain your thinking.

Math @ Home Activity

Ask your child to explain the patterns they see when adding 10 or 100 to a 3-digit number. Have him or her write different equations to show the patterns. Give him or her different colored crayons or pencils to circle the number that changes in each sum. Cut the equations out and have him or her organize them in piles of add 10 and add 100.

Additional Practice

Name _____

Review

To add 3-digit numbers, you can add the ones, then the tens, and finally the hundreds.

132 + 154 = ?

You can use base-ten shorthand to represent 3-digit addition problems.

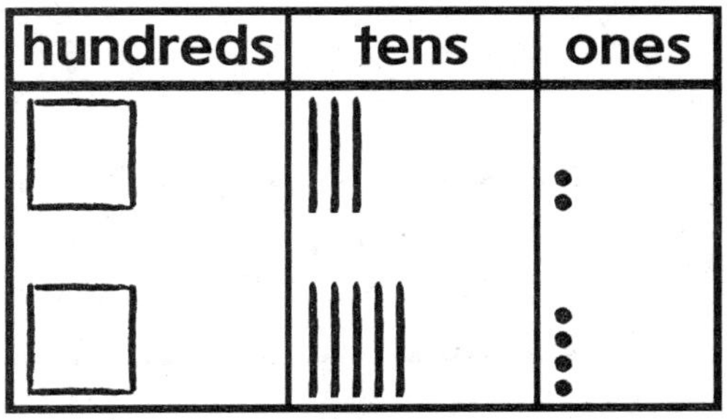

hundreds	tens	ones
☐	|||	⋮
☐	|||||	⋮

132 + 154 = 286

Is the statement true or false? Circle the correct answer.

1. The total number of tens in the sum of 51 + 235 is 6.

True   False

2. The total number of hundreds in the sum of 327 + 462 is 7.

True   False

What is the sum? Use base-ten shorthand to show your work.

3. 67 + 221 = _____

hundreds	tens	ones

4. 145 + 334 = _____

hundreds	tens	ones

5. DeShawn works at a book store. He sells 184 books on Saturday and 212 books on Sunday. Deshawn says he sold 386 books on Saturday and Sunday. How do you respond to Deshawn?

Math @ Home Activity

Draw and cut out base-ten blocks from pieces of paper. On a sheet of paper, write an addition equation that involves adding two 3-digit numbers without regrouping. Have your child create two groups of base-ten blocks to represent adding the two 3-digit numbers in the equation. Have your child find the sum. Repeat with a different addition equation.

Additional Practice

Name _____

Review

You can use base-ten blocks to represent 3-digit addition problems.

Tori's mystery book has 248 pages and her animal book has 166 pages. How many pages do the 2 books have?

$$248 + 166 = 414$$

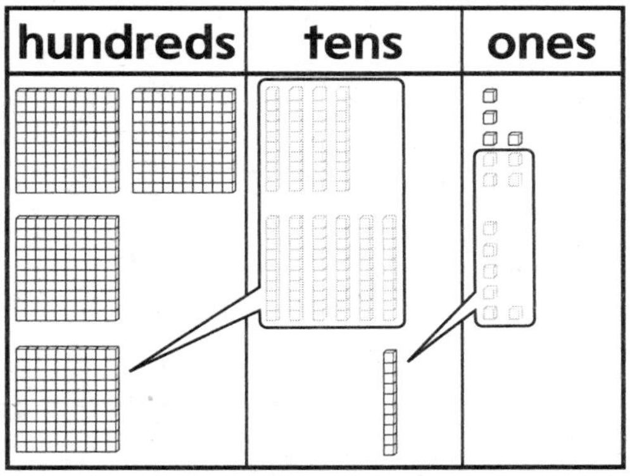

The mystery and animal books have 414 pages.

I. Which equations need regrouping? Choose all the correct answers.

A. $137 + 321 = ?$ **B.** $204 + 458 = ?$

C. $563 + 291 = ?$ **D.** $344 + 635 = ?$

What is the sum? Use base-ten shorthand to show your work.

2. 257 + 118 = _____

3. 336 + 295 = _____

---

4. Ian is playing a game. He earns 383 points in the first round and 549 points in the second round. He needs 925 points or more to win the game. Does Ian have enough points to win? Explain your thinking.

Math @ Home Activity

Write an equation on a sheet of paper that involves adding two 3-digit numbers with regrouping. Have your child draw a place-value chart like the one shown in Exercise 3. Then tell your child to draw base-ten shorthand in the place-value chart to represent and solve the equation. Have your child explain each step to you as he or she finds the sum.

Additional Practice

Name _____

Review

You can add 3-digit numbers by decomposing both addends by place value to find partial sums. Then add the partial sums.

$$152 \quad + \quad 236 \quad = ?$$

$100+50+2 \quad 200+30+6$

$100 + 200 = 300$
$50 + 30 = 80$
$2 + 6 = 8$
$300 + 80 + 8 = 388$
So, $152 + 236 = 388$.

$$397 \quad + \quad 465 \quad = ?$$

$300+90+7 \quad 400+60+5$

$300 + 400 = 700$
$90 + 60 = 150$
$7 + 5 = 12$
$700 + 150 + 12 = 862$
So, $397 + 465 = 862$.

I. Which equation shows both addends decomposed by place value?

A. $114 \quad + \quad 279 \quad = ?$

$100 + 10 + 4 \quad 200 + 7 + 9$

B. $483 \quad + \quad 321 \quad = ?$

$400+80+3 \quad 300+20+1$

2. What is the sum? Decompose both addends to solve.

a.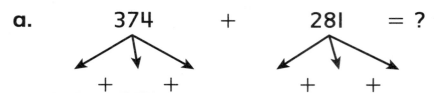

374   +   281   = ?

___ + ___ + ___   ___ + ___ + ___

b. Add hundreds: _____ + _____ = _____

Add tens: _____ + _____ = _____

Add ones: _____ + _____ = _____

c. Solve using partial sums:

_____ + _____ + _____ = _____

What is the sum? Decompose both addends to solve.

3. 423 + 249 = _____   4. 764 + 158 = _____

---

5. A grocery store has 296 red apples and 362 green apples. How many apples are there in all? Decompose both addends to solve.

Additional Practice

Name _____

Review

You can decompose one addend and then add
by place value to add 3-digit numbers.

One Way

375 + 229

200 + 20 + 9

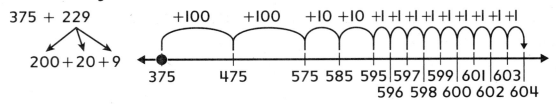

Another Way

375 + 229

25 + 200 + 4

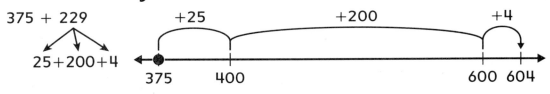

How can you decompose one addend? Choose all
the correct answers.

1. 146 + 459 = ?

    A. 100 + 40 + 6 + 459   **B.** 100 + 4 + 6 + 459

    C. 146 + 400 + 50 + 9   **D.** 146 + 400 + 5 + 9

2. 238 + 615 = ?

    A. 238 + 600 + 1 + 5   **B.** 238 + 600 + 10 + 5

    C. 200 + 3 + 8 + 615   **D.** 200 + 30 + 8 + 615

3. What is the sum? Decompose one addend to solve.

   a. 501 + 382 = ?

         ____ + ____ + ____

   b. Add: _____ + _____ + 80 + 2 = _____

   c. Solve: 501 + 382 = _____

What is the sum? Decompose one addend. Use a number line to show your work.

4. 463 + 247 = _____

5. 715 + 124 = _____

---

6. Carmen is finding 379 + 221. She decomposes 221 as 200 + 2 + 1 while finding the sum. Carmen says the sum is 582. How do you respond to Carmen?

Math @ Home Activity

Write an addition equation in which there are two 3-digit addends. Then create number tiles with paper that your child can use to find the sum. Make enough number tiles so your child can decompose the lesser addend while finding the sum. Check his or her work before repeating this activity with a new addition equation.

Additional Practice

Name _____

Review

You can adjust addends to make 3-digit addends friendlier to add.

Adjust the addends to solve $496 + 319$.

You can adjust either addend first.

What is added to one addend must be subtracted from the other addend.

$496 + 319 = 815$

$\boxed{+4}\quad\boxed{-4}$

$500 + 315 = 815$

$496 + 319 = 815$

$\boxed{-1}\quad\boxed{+1}$

$495 + 320 = 815$

I. How can you adjust the addends? Choose all the correct answers.

$$277 + 398 = \underline{\hphantom{0000}}$$

A. $280 + 395$ **B.** $280 + 401$

C. $275 + 400$ **D.** $275 + 396$

How can you adjust addends to find the sum? Fill in the numbers.

2. 186 + 297

☐ ☐
↓ ↓

____ + ____ = ____

3. 578 + 404

☐ ☐
↓ ↓

____ + ____ = ____

4. 395 + 528

☐ ☐
↓ ↓

____ + ____ = ____

5. 693 + 199

☐ ☐
↓ ↓

____ + ____ = ____

---

6. What is the sum? Adjust the addends. Use a number line to show your work.

503 + 364

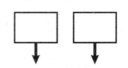

Additional Practice

Name _____

Review

You can use different addition strategies to add 3-digit numbers. The sum will be the same.

Decompose Addends

397   +   234   = ? $300 + 200 = 500$
                            $90 + 30 = 120$
$300+90+7$ $200+30+4$ $7 + 4 = 11$
                            $500 + 120 + 11 = 631$

Adjust Addends

$397 + 234 = 631$

$\boxed{+3}$ $\boxed{-3}$

$400 + 231 = 631$

---

1. What addition strategy is shown? Choose the correct answer.

184   +   429   = ? $100 + 400 = 500$
                              $80 + 20 = 100$
$100 + 80 + 4$ $400+20+9$ $4 + 9 = 13$
                              $500 + 100 + 13 = 613$

A. adjust addends   **B.** decompose both addends

C. decompose one addend   **D.** skip counting

2. What addition strategy is shown? Choose the correct answer.

$$602 + 289 = 891$$

-2	$+2$
↓	↓

$$600 + 291 = 891$$

A. adjust addends

B. decompose both addends

C. decompose one addend

D. skip counting

---

3. A store has 358 red apples and 273 green apples. How many red and green apples are there?

4. There were 198 people attending a school play on Friday night and 246 people attending on Saturday night. How many people attended the play on the two nights? Explain what strategy you used and why.

Math @ Home Activity

Write an addition equation in which there are two 3-digit addends. Have your child use one of the strategies from this lesson to find the sum. Have your child explain why he or she chose the strategy he or she used to find the sum. Repeat and have your child use a different strategy to find the sum this time.

Additional Practice

Name _____

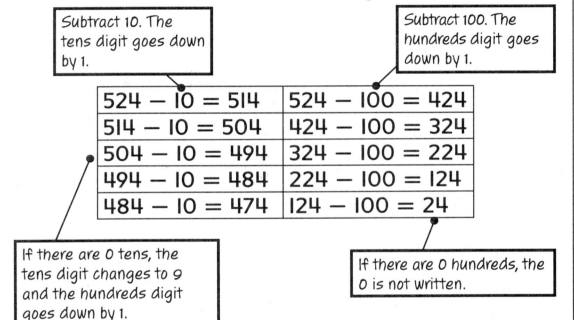

Review

You can use patterns to subtract 10 and 100 from 3-digit numbers.

The table shows patterns that take place when 10 and 100 are subtracted from 3-digit numbers.

Subtract 10. The tens digit goes down by 1.

Subtract 100. The hundreds digit goes down by 1.

$524 - 10 = 514$	$524 - 100 = 424$
$514 - 10 = 504$	$424 - 100 = 324$
$504 - 10 = 494$	$324 - 100 = 224$
$494 - 10 = 484$	$224 - 100 = 124$
$484 - 10 = 474$	$124 - 100 = 24$

If there are 0 tens, the tens digit changes to 9 and the hundreds digit goes down by 1.

If there are 0 hundreds, the 0 is not written.

1. Which equations are true? Choose all the correct answers.

   A. $300 - 10 = 20$ **B.** $300 - 10 = 290$

   C. $300 - 100 = 200$ **D.** $300 - 100 = 400$

What is the difference? Use the number line.

2. 206 − 10 = _____

3. 153 − 100 = _____

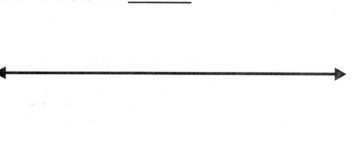

What is the difference? Use patterns to help you.

4. 351 − 10 = _____   **5.** 619 − 10 = _____

6. 185 − 100 = _____   **7.** 902 − 100 = _____

8. James has 138 cards. Luis has 10 fewer cards than James. How many cards does Luis have?

9. Maxine has 459 stamps. She sells 100 stamps. How many stamps are left?

Math @ Home Activity

Create a subtraction problem based on a situation in your home that requires subtracting 10 or 100. Ask your child to explain how he or she can find the difference in his or her head. Take turns with your child creating problems and subtracting 10 and 100 in your head.

Additional Practice

Name _____

Review

You can use base-ten blocks to represent and solve 3-digit subtraction equations.

Find 354 − 121.

Represent the problem with an equation and base-ten blocks.

Start with 354.

Take away 121.

233 is the difference.

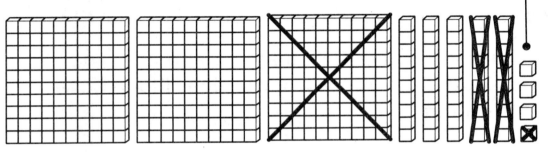

$$354 - 121 = 233$$

---

1. Which equation is represented by the base-ten blocks? Choose the correct answer.

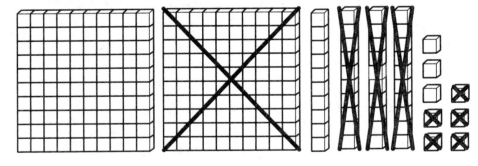

A. 248 − 125 = 123   **B.** 245 − 135 = 110

C. 248 + 135 = 113   **D.** 248 − 135 = 113

What is the difference? Use base-ten shorthand to show your work.

2. $276 - 142 =$ _____

3. $367 - 153 =$ _____

Represent the problem using base-ten shorthand.

4. Elena's book has 289 pages. Elena has read 134 pages. How many pages does Elana have left to read?

5. Mario has 348 pictures on his camera. He deletes 126 of the pictures. How many pictures are left on his camera?

Math @ Home Activity

With your child, draw and cut out base-ten blocks. Make 3 hundreds flats, 9 tens rods, and 9 ones units. Write a subtraction problem. Have your child move base-ten blocks away from the group to find the difference. Repeat with a different problem.

Additional Practice

Name _____

Review

You can decompose one number and count back on a number line to subtract 3-digit numbers.

Find 528 − 349.

You can decompose by place value.

528 − 349 = 179

300 + 40 + 9

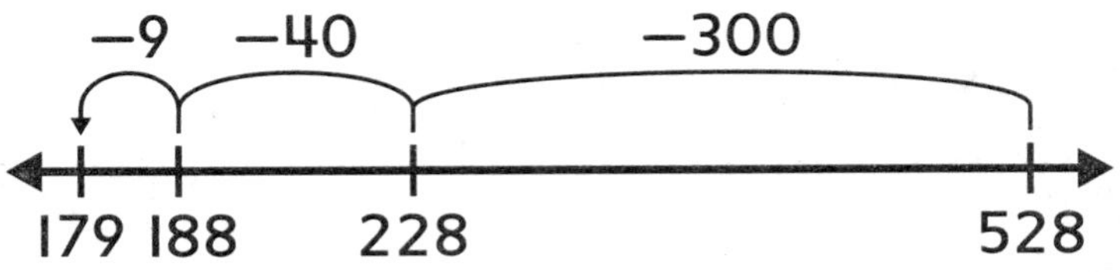

−9 −40   −300

179 188   228   528

You can decompose a different way.

528 − 349 = 179

328 + 20 + 1

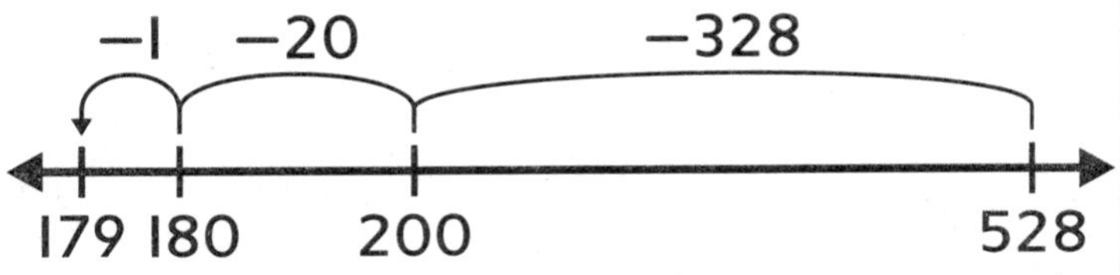

−1 −20   −328

179 180   200   528

How can you decompose the change number?
Circle the correct answer.

1. 441 − 156 = ?

100 + 5 + 6

100 + 50 + 6

2. 613 − 234 = ?

213 + 20 + 1

200 + 3 + 40

3. 702 − 467 = ?

400 + 60 + 7

46 + 7

4. 875 − 690 = ?

600 + 90

600 + 9

How can you decompose to find the difference? Show the subtraction on the number line.

5. $516 - 228 =$ _____

   $228 =$ _____ $+$ _____ $+$ _____

6. $804 - 439 =$ _____

   $439 =$ _____ $+$ _____ $+$ _____

7. Gil decomposes by place value to find the difference of $635 - 342$. What other way can he decompose to find the difference?

Math @ Home Activity

Write a 3-digit subtraction problem that requires regrouping on a sheet of paper. Have your child show you two ways to decompose to find the difference. Then have your child explain which way is more efficient for the equation.

Additional Practice

Name _____

Review

You can count on using a number line to solve subtraction equations with 3-digit numbers.

Find 719 − 486.

Represent the problem on a number line.

| Start at one number and count on to the other number. |

| The total of the jumps is the difference. |

233

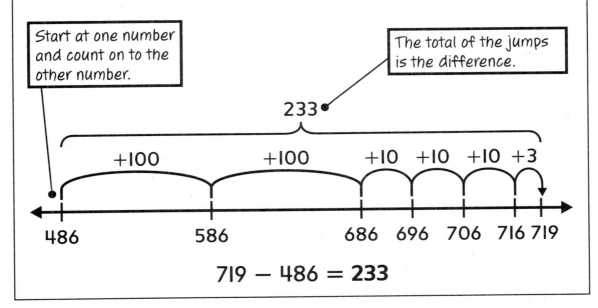

+100   +100   +10 +10 +10 +3

486   586   686 696 706 716 719

719 − 486 = **233**

---

1. How can you count on to subtract? Fill in the numbers on the number line and find the difference.

567 − 323 = _____

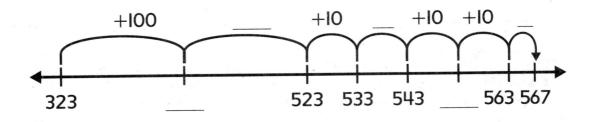

+100   ___   +10   ___ +10 +10   ___

323   ___   523 533 543 ___ 563 567

2. Which equation is related to $641 - 259 = ?$
   Choose the correct answer.

    A. $259 + 641 = ?$ **B.** $? - 259 = 641$

    C. $259 + ? = 641$ **D.** $641 + 259 = ?$

What is the difference? Use the number line to count on.

3. $492 - 275 = $ _____

275   492

4. $844 - 518 = $ _____

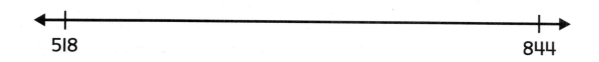

518   844

---

5. There are 466 bottles of water for sale at a grocery store. If 137 bottles are sold, how many bottles of water are left? Explain your thinking.

Math @ Home Activity

Write a problem that involves subtracting two 3-digit numbers that resembles a situation in your child's life. Then have him or her solve the problem using a number line to count on. Have your child explain how he or she found the difference.

Additional Practice

Name _____

Review

You can regroup a ten if needed when subtracting 3-digit numbers.

$431 - 215 = ?$

Represent the problem with base-ten blocks.

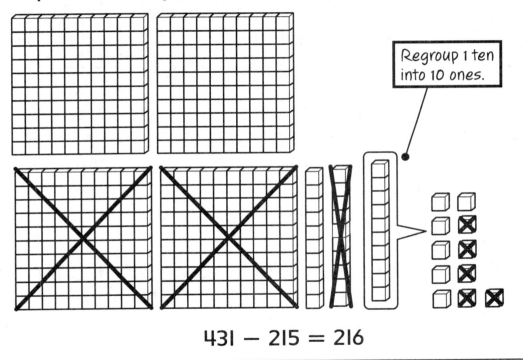

Regroup 1 ten into 10 ones.

$431 - 215 = 216$

Is regrouping needed to subtract?
Choose Yes or No.

1. $346 - 123$

   Yes   No

2. $552 - 237$

   Yes   No

3. $891 - 674$

   Yes   No

4. $968 - 716$

   Yes   No

What is the difference? Show the subtraction on the base-ten blocks.

5. 253 − 126 = _____

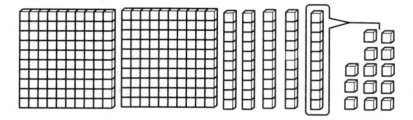

6. 371 − 264 = _____

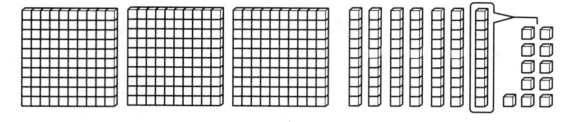

---

7. a. There are 534 people at a school picnic. Then 328 people leave. How many people are still at the picnic?

b. Explain why regrouping is needed to find the number of people that are still at the picnic.

Math @ Home Activity

Write a subtraction problem that involves 3-digit numbers on a sheet of paper. Have your child draw base-ten blocks and cross some of them out to find the difference. Ask your child to explain how he or she used the base-ten blocks to subtract the 3-digit numbers. Repeat with different subtraction problems.

Additional Practice

Name _____

Review

You can regroup a hundred and a ten if needed when subtracting 3-digit numbers.

$300 - 136 = ?$

Represent the problem with base-ten blocks.

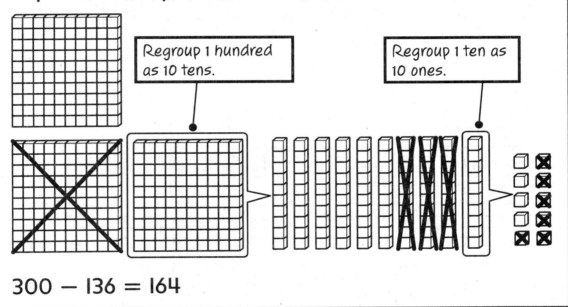

Regroup 1 hundred as 10 tens.

Regroup 1 ten as 10 ones.

$300 - 136 = 164$

What needs to be regrouped to subtract? Circle the correct answer.

1. $428 - 249$

tens   hundreds   both

2. $754 - 517$

tens   hundreds   both

What is the difference? Show the subtraction on the base-ten blocks.

3. 363 − 172 = _____

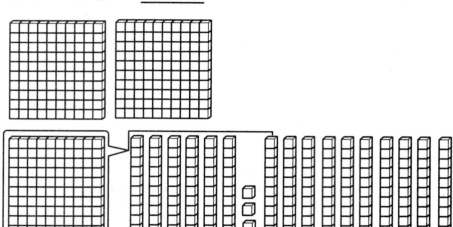

4. 441 − 256 = _____

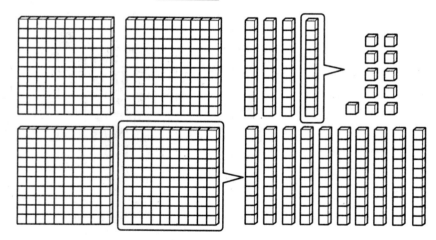

Math @ Home Activity

Write a subtraction problem that involves 3-digit numbers and requires regrouping of one place value on a sheet of paper. Have your child identify which place value needs to be regrouped to subtract. Then have your child draw base-ten blocks to help him or her find the difference. Repeat with another subtraction problem that requires regrouping of the tens and hundreds.

Additional Practice

Name _____

Review

You can adjust 3-digit numbers to make them friendlier to subtract.

On Monday, Juan has 347 e-mails. He deletes 196 e-mails. On Tuesday, he has 442 e-mails and deletes 235 of them. How many e-mails does he have left each day?

Adjust both numbers in the equation by the same amount.

$347 - 196 = 151$ $\boxed{+4}$ $\boxed{+4}$ $\downarrow$ $\downarrow$ $351 - 200 = 151$	$442 - 235 = 207$ $\boxed{-5}$ $\boxed{-5}$ $\downarrow$ $\downarrow$ $437 - 230 = 207$
Juan has 151 e-mails left on Monday.	Juan has 207 e-mails left on Tuesday.

---

1. How can you adjust the numbers to subtract? Choose all the correct answers.

$$553 - 248 = ?$$

A. $550 - 251$ **B.** $550 - 245$

C. $551 - 250$ **D.** $555 - 250$

How can you adjust to find the difference? Fill in the numbers.

2. $684 - 397 =$ _____

```
[    ]  [    ]
   ↓       ↓
```
_____ − _____ = _____

3. $871 - 462 =$ _____

```
[    ]  [    ]
   ↓       ↓
```
_____ − _____ = _____

---

4. Dane has 469 papers in a box. He removes 205 papers from the box. How many papers are still in the box? What friendly equation can you use to solve?

5. In March, 763 cans of food are donated to a food bank. In April, 554 cans are donated to the food bank. How many more cans of food are donated in March than in April? Use a friendly equation to solve the problem. Explain your thinking.

Math @ Home Activity

Write a subtraction problem that involves two 3-digit numbers on a piece of paper. Suggest an adjustment for the problem to make a friendlier subtraction problem. Ask your child to determine if the suggested adjustment will work. Have him or her show their check of the adjustment. Have him or her carry out the subtraction to find the difference. Repeat with other subtraction problems.

Additional Practice

Name _____

Review

You can use different subtraction strategies to subtract 3-digit numbers. The difference will stay the same no matter what strategy is used.

Find $673 - 498$.

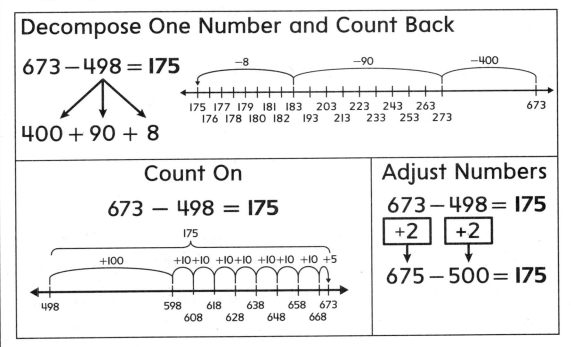

Fill in the correct answer to complete the sentence.

1. To count on to find the difference of $365 - 127$, start at _____.

2. To count back to find the difference of $534 - 219$, start at _____.

---

3. How can you adjust the numbers to find the difference? Choose all the correct answers.

$447 - 254 = ?$

A. $450 - 257$

B. $450 - 251$

C. $443 - 250$

D. $451 - 250$

4. How can you decompose 336 to find the difference of $751 - 336$? Choose all the correct answers.

A. $33 + 6$

B. $300 + 30 + 6$

C. $3 + 3 + 6$

D. $300 + 30 + 5 + 1$

5. What equation is related to the equation $? = 842 - 577$? Choose all the correct answers.

A. $842 + 577 = ?$

B. $? - 577 = 842$

C. $577 + ? = 842$

D. $842 - ? = 577$

---

6. Use a subtraction strategy to solve the problem below. Explain the subtraction strategy you used.

Chase helps count birds at a nature park. In May, he counts 418 birds. In June, he counts 621 birds. How many more birds does Chase count in June than in May?

Math @ Home Activity

Write a 3-digit subtraction problem that requires regrouping on a sheet of paper. Have your child solve the problem in three different ways by using the three strategies reviewed in this lesson. Then have your child identify the strategy he or she prefers and explain why.

Additional Practice

Name _____

Review

You can use addition and subtraction strategies to solve one- and two-step word problems.

Fern has 400 beads. She gives 100 beads to her sister. Then she uses 128 beads to make necklaces. How many beads does Fern have now?

First, make sense of the problem and represent it.

|-------- 400 beads --------|

100	128	?

Then, choose a strategy to solve the problem.

One Way

$100 + 128 = 228$

$400 - 228 = 172$

Fern has 172 beads now.

Another Way

$400 - 128 = 272$

$272 - 100 = 172$

I. Which equation can you use to represent the word problem? Circle the correct answer.

There are 576 mammals and 239 reptiles at a zoo. How many more mammals than reptiles are at the zoo?

A. $576 + 239 = ?$ **B.** $576 - 239 = ?$

C. $239 + 576 = ?$ **D.** $? - 239 = 576$

Write an equation to represent the problem. Use any strategy to solve.

2. There are 293 posters in a school. Then 165 of the posters are taken down and 112 new posters are put up. How many posters are in the school now?

3. There are 371 birds at a park. There are 144 fewer squirrels than birds at the park. How many birds and squirrels are at the park?

4. Carmen wins 487 tickets while playing games at an arcade. Then she wins another 136 tickets. She uses 225 tickets to get a prize. How many tickets does she have now? Explain how you solved the problem.

Additional Practice

Name

Review

A picture graph uses pictures to represent data. You can draw a picture graph to organize and interpret data.

Rubie records each student's favorite toy. How can you represent this information in a different way?

Favorite Toy	
Toy	**Tally**
Airplane	ЖЖ |||
Doll	ЖЖ ||||
Teddy Bear	ЖЖ ||
Puppet	ЖЖ

Draw a picture graph to represent this data.

title

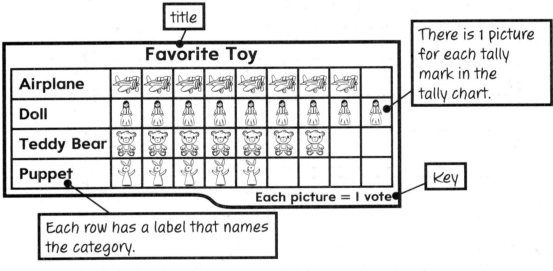

There is 1 picture for each tally mark in the tally chart.

Favorite Toy

Each picture = 1 vote

Key

Each row has a label that names the category.

---

1. Use the picture graph to answer the questions.

Favorite Animals at the Aquarium

Shark	🦈 🦈 🦈 🦈 🦈 🦈 🦈	
Lobster	🦞 🦞 🦞 🦞 🦞	
Clown fish	🐠 🐠 🐠 🐠	
Starfish	⭐ ⭐ ⭐ ⭐ ⭐ ⭐ ⭐ ⭐	

Each picture = 1 vote

a. How many people voted for sharks?

b. How many people voted for lobsters?

c. Which animal got the fewest votes?

d. Which animal got the most votes?

e. How many people were surveyed? Explain.

2. Molly read 5 books, Sal read 8 books, and Xin read 1 less book than Sal. How can you represent this data using a picture graph?

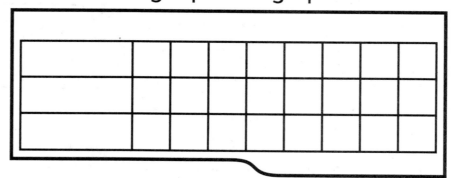

Math @ Home Activity

Have your child ask friends and family a survey question and record the answers in a tally chart. Then help your child create a picture graph to represent the data. Ask him or her questions about the data.

Additional Practice

Name _____

Review

A bar graph uses bars to show data. You can use bar graphs to organize and interpret data.

An art teacher made a tally chart to record the colors her students chose for an art project. What other ways could you represent the data?

Crayons Chosen for Art Project	
Color	**Tally**
Red	卌 IIII
Blue	III
Yellow	卌 II
Green	卌

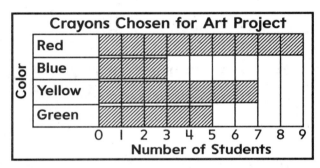

I. How can you represent the data using a horizontal bar graph?

Favorite Music	
Type of Music	**Tally**
Classical	IIII
Rock	卌
Country	II
Pop	卌 I

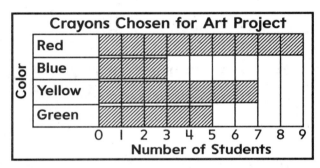

2. Use the data to answer the questions.

a. Dane's class voted for their favorite fall activity. Each student voted once. How can you represent the data using a vertical bar graph?

Favorite Fall Activity	
Activity	**Tally**
Carving Pumpkins |卌 |||
Picking Apples	卌
Jumping in Leaves	卌	
Corn Maze	卌 ||

Favorite Fall Activity

Number of Students

9
8
7
6
5
4
3
2
1
0

Carving Pumpkins   Picking Apples   Jumping in Leaves   Corn Maze

Activity

b. How many students chose jumping in leaves as their favorite activity?

c. What activity was chosen the most?

d. How many students chose a corn maze as their favorite activity?

e. What activity was chosen the least?

Math @ Home Activity

Have your child ask family members to respond to a survey question. If weather permits, have your child create a bar graph for the data using sidewalk chalk on the sidewalk or driveway. Have your child explain their graph. If going outside is not an option, have your child create the graph on paper.

Additional Practice

Name _____

Review

A bar graph can be a helpful tool for solving problems with data.

You can add and subtract to make observations about the data.

What are some ways you can add, subtract, and compare the data in the bar graph?

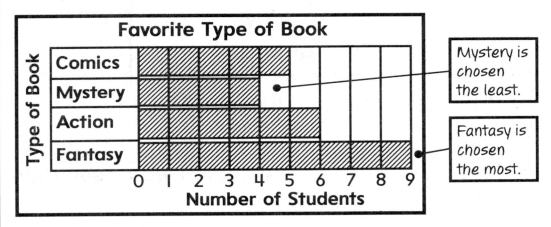

3 more people chose fantasy than action. $9 - 6 = 3$

9 people chose comics and mystery. $5 + 4 = 9$

5 more people chose fantasy than mystery. $9 - 4 = 5$

---

1. Use the bar graph above to answer the questions.

   a. How many more people chose action than mystery?

   b. How many fewer people chose comics than fantasy?

2. Use the bar graph to answer the questions.

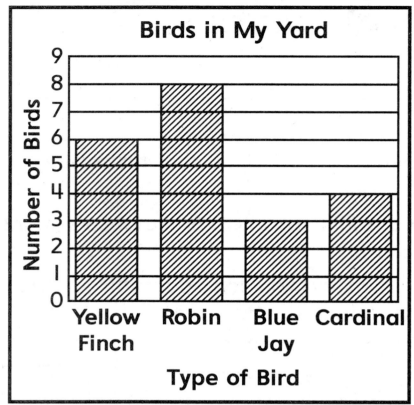

a. What bird was seen the most?

b. How many more yellow finches were seen than blue jays?

c. How many fewer cardinals were seen than robins?

d. How many birds were seen in all? Explain.

Help your child create his or her own bar graph about birds seen outside your home. Then ask your child questions about the data in the graph.

Additional Practice

Name _____

Review

Measurement data can be collected by measuring and recording the length of objects. The data can be organized in a tally chart.

Selena measured 8 crayons in centimeters. How can you organize the data?

Make 1 tally mark in the tally chart to represent each piece of data.

| 8 centimeters |
| 7 centimeters |
| 5 centimeters |
| 6 centimeters |
| 8 centimeters |
| 7 centimeters |
| 6 centimeters |
| 8 centimeters |

Length of Crayon	
Length (centimeters)	Tally
5	|
6	||
7	||
8	|||

---

1. How can you make a tally chart to show the data?

| 8 inches |
| 10 inches |
| 8 inches |
| 7 inches |
| 8 inches |
| 10 inches |
| 7 inches |
| 8 inches |

Length of Paintbrush	
Length (inches)	Tally
7	
8	
9	
10	

2. Measure the lengths of 8 pencil boxes. Collect the data in a list. Then make a tally chart to show the data.

3. Use the data to answer the questions.

a. Harry is making a tally chart to show this data. How many rows should his tally chart have? Explain your thinking.

17 inches
19 inches
16 inches
21 inches
20 inches
18 inches
19 inches
16 inches
18 inches
21 inches

b. How many tally marks go in the row for 18 inches?

c. How would the tally chart change if Harry's list had a measurement of 22 inches?

Math @ Home Activity

Have your child measure the lengths of 8 household objects in centimeters. Tell your child to record the measurements in a list. Then have your child use the list of measurements to create a tally chart.

Additional Practice

Name _____

Review

A line plot is a graph with Xs above a number line.

Sue measured the lengths of necklaces.

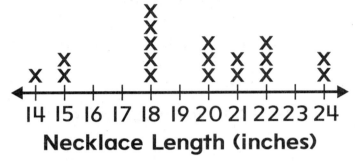

Necklace Length (inches)

The most common length is 18 inches.

The least common length is 14 inches.

$1 + 2 + 5 + 3 + 2 + 3 + 2 = 18$ necklaces measured

I. Ian measured the lengths of zucchinis in his garden.
Use the data on the line plot to answer the questions.

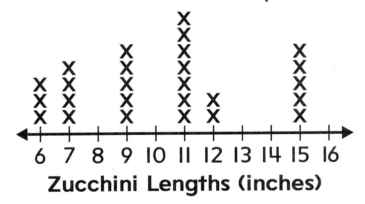

Zucchini Lengths (inches)

a. What is the length of the shortest zucchini?

b. How many measurements were recorded?

2. Sara measured the lengths of the shells she collected. Use the data on the line plot to answer the questions.

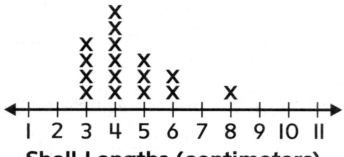

Shell Lengths (centimeters)

a. What is the most common length measured?

b. What is the least common length measured?

c. What is the length of the longest shell?

d. What is the length of the shortest shell?

e. How many measurements were recorded?

_____

3. What is an advantage of using a line plot to display data?

_____

Additional Practice

Name _____

Review

You can use a line plot to make observations about data.

Each X on a line plot represents one value in a set of data.

Mrs. Ward's class measured the lengths of leaves. What length is the most common?

Draw an X for each measurement in the list.

11 centimeters
13 centimeters
12 centimeters
11 centimeters
10 centimeters
11 centimeters
13 centimeters
12 centimeters
10 centimeters
11 centimeters
13 centimeters
11 centimeters
13 centimeters
12 centimeters

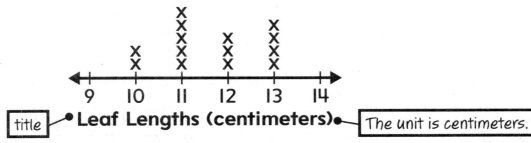

title • **Leaf Lengths (centimeters)** • The unit is centimeters.

11 centimeters is the most common length.

1. How can you represent the data using a line plot?

a. Students in Mr. Hale's class measured the lengths of paper chains they created. Use the data to make a line plot.

55 inches
52 inches
54 inches
55 inches
58 inches
56 inches
55 inches
58 inches

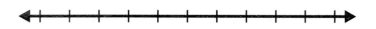

b. Write a question that can be answered by looking at the line plot.

Additional Practice

Name _____

Review

You can identify 2-dimensional shapes by the number of sides, angles, or vertices.

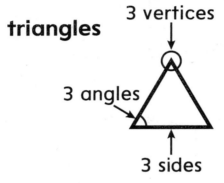

triangles

3 vertices

3 angles

3 sides

quadrilaterals

4 vertices

4 angles

4 sides

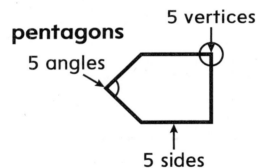

pentagons

5 vertices

5 angles

5 sides

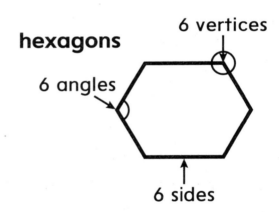

hexagons

6 vertices

6 angles

6 sides

How many sides, angles, and vertices does the shape have?

1.

_____ sides

_____ angles

_____ vertices

2.

_____ sides

_____ angles

_____ vertices

How many sides, angles, and vertices does the shape have?

3.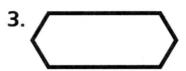

_____ sides

_____ angles

_____ vertices

4.

_____ sides

_____ angles

_____ vertices

5. Which shapes are pentagons? Choose all the correct answers.

A.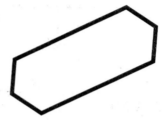

B.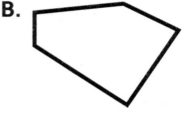

C.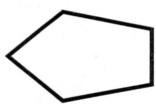

D.

6. Jeremy has a pet fish. What shape is the side of the fish tank? Explain how you know.

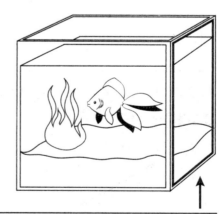

Math @ Home Activity

Have your child identify objects around your home that are shaped like triangles, quadrilaterals, pentagons, and hexagons. Ask your child to identify the number of sides, number of angles, number of vertices, and the name of each shape.

Additional Practice

Name _____

Review

You can draw a 2-dimensional shape when given its attributes.

3 sides and 3 angles

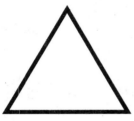

4 sides, 4 angles, and all sides the same length

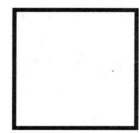

4 sides, 4 angles, and opposite sides the same length

5 sides, 5 angles, and all sides the same length

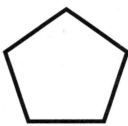

6 sides, 6 angles, and all sides different lengths

I. What shape has 4 sides, 4 angles, and all sides different lengths? Draw the shape. Then write its name.

Draw the shape. Then write the name.

2. What shape has 6 sides, 6 angles, and all sides different lengths?

3. What shape has 3 sides, 3 angles, and all sides the same length?

4. What shape has 5 sides, 5 angles, and all sides the same length?

5. a.   Katie says she drew a rectangle. How do you respond to her?

b.   What are 3 attributes of the shape Katie drew?

Math @ Home Activity

Ask your child to draw a triangle, quadrilateral, pentagon, and hexagon. Have your child explain the attributes of each shape. Then challenge your child by asking him or her to draw two more different examples of each shape.

Lesson 12-3

Additional Practice

Name ______________________________

Review

You can recognize 3-dimensional shapes by their attributes. The number of faces, edges, and vertices can help you identify solids.

	Cone	Cylinder	Sphere	Rectangular Prism
Face or Base	1 circle	2 circles	0	6 rectangles
Edge	0	0	0	12
Vertex or Apex	1	0	0	8
Example	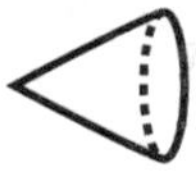	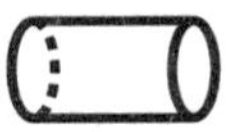	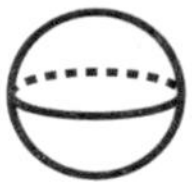	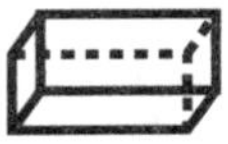

How many faces, edges, and vertices does the shape have? What is the shape?

1. 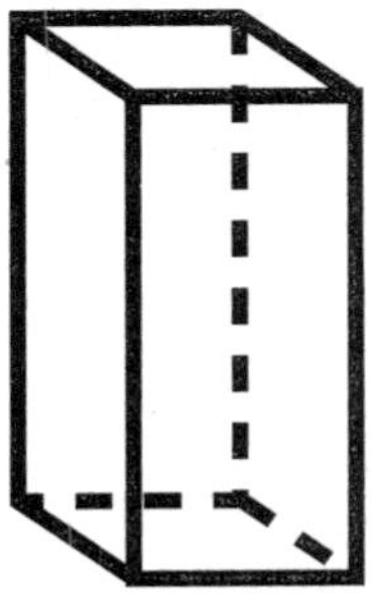

_____ faces

_____ edges

_____ vertices

This shape is a

_______________.

2. 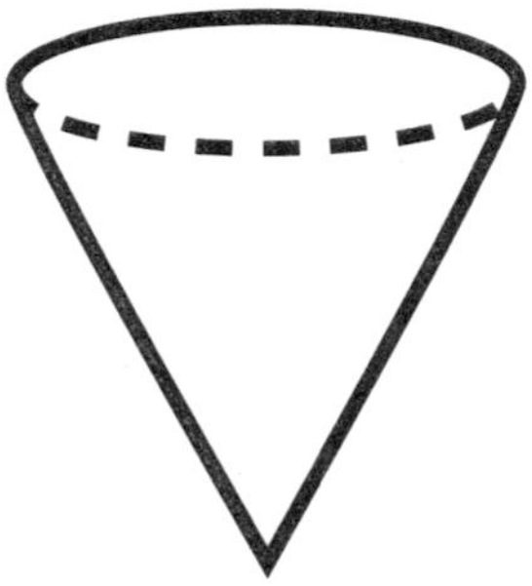

_____ base

_____ edges

_____ apex

This shape is a

_______________.

3. Which shapes are cubes? Choose all the correct answers.

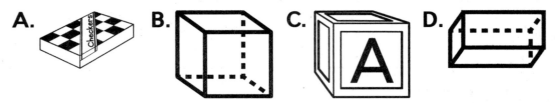

A.   B.   C.   D.

4. Which shapes are cylinders? Choose all the correct answers.

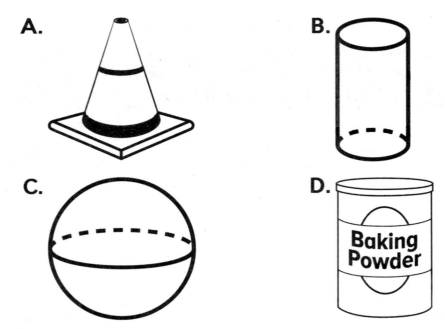

A.

B.

C.

D. Baking Powder

5. Kendra has an object with 0 faces, 0 edges, and 0 vertices. What shape is the object? Explain.

Math @ Home Activity

Have your child participate in a 3-dimensional shape scavenger hunt. Tell your child the name of a 3-dimensional shape. Then have your child look around your home for an object that is that shape and bring it to you. Ask your child to explain how many faces, edges, and vertices the object has. Repeat until your child has found an object for each 3-dimensional shape.

Additional Practice

Name _____

Review

You can partition shapes, such as circles and rectangles, into equal shares. Equal shares are shares that are the same size.

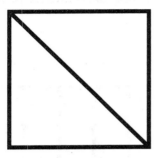  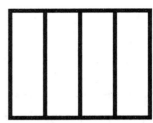

2 halves   3 thirds   4 fourths

Which shapes are partitioned into equal shares? Choose all the correct answers.

1. A. 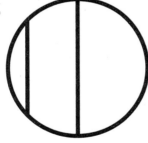   B. 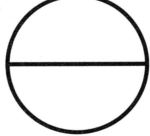   C.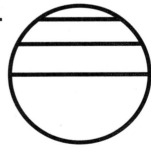

2. A. 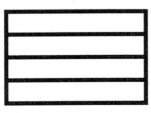   B. 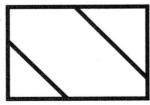   C.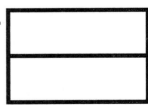

3. How can you partition the circle into 4 equal shares? Draw to show your work.

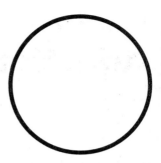

4. How can you partition the rectangle into 2 equal shares? Draw to show your work.

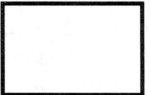

5. Dylan has a square piece of fabric. She says she partitioned the fabric into 3 equal shares. How do you respond to Dylan?

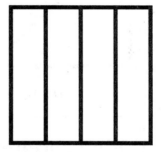

Math @ Home Activity

Cut out several circles and rectangles from construction paper. Fold some of the shapes into equal and unequal halves, thirds, and fourths. Have your child identify the shapes that are partitioned into equal shares. Then give him or her shapes that are not folded and have him or her fold them into halves, thirds, and fourths.

Additional Practice

Name

Review

You can partition shapes, such as circles or rectangles, into equal shares in different ways. The equal shares do not have to be the same shape.

halves   thirds   fourths

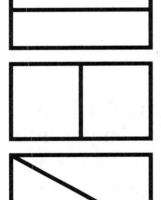
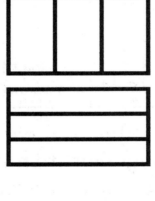

I. Which shows how to partition the same square into halves? Choose all the correct answers.

A.    B. 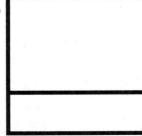   C.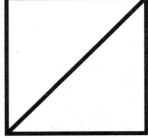

2. Which shows how to partition the same circle into thirds? Choose all the correct answers.

A.  B. 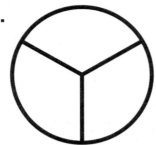 C.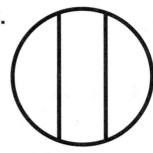

3. How can you partition the square into fourths? Show two different ways.

4. Lulu wants to partition the two rectangles into thirds to make flags. Partition the rectangles into thirds. Then explain why the two rectangles can be partitioned into thirds in two different ways.

Math @ Home Activity

Ask your child to define halves, thirds, and fourths. Give him or her a circular slice of an apple. Have your child show, with his or her finger, how to partition the apple slice into halves, thirds, and fourths. Ask your child if there is more than one way to partition the apple slice into halves, thirds, and fourths.

Additional Practice

Name _____

Review

You can partition rectangles into rows and columns using squares of equal size. Then you can use repeated addition to find the total number of squares in the rows and columns.

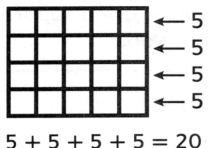

$5 + 5 + 5 + 5 = 20$

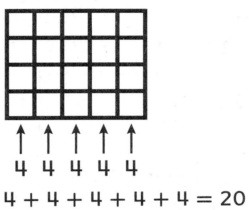

$4 + 4 + 4 + 4 + 4 = 20$

How many rows, columns, and squares of equal size is the rectangle partitioned into? Write an equation to find the total number of squares.

1.
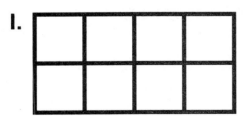

a. Rows: _____

b. Columns: _____

c. Equation: _____

d. Total squares: _____

2.

a. Rows: _____

b. Columns: _____

c. Equation: _____

d. Total squares: _____

How can you partition the rectangle using squares of equal size? Draw to show your work.

3. ▢

Total squares: _____

4. ▢

Total squares: _____

5. A rectangular board game is divided into 5 rows and 6 columns of squares. How many squares are on the board? Explain your reasoning.

Cut out 30 small squares that are the same size. Have your child build a rectangle with some of the squares. Have him or her identify the number of squares in each row and each column. Then have your child write an equation to find the total number of squares. Repeat this activity several times.